Life After 100,000 Miles

Author

Theodore R. Hansen

Life After 100,000 Miles

Author: Theodore R. Hansen

First Edition

Dedication

I dedicate *Life After 100,000 Miles* to my two sons, Samuel V. Hansen and David T. Hansen, who inspired me to write this book. They have a great interest in cars, but they do not have a complete mechanical knowledge of how cars operate. They also wanted to know how to do some of the car maintenance themselves. It is with great joy I write this book for my sons and the readers of this book. You can learn and apply this knowledge to maintain your own vehicles.

Acknowledgments

I would like to thank the following people, companies, and the Versailles, Indiana State Police Post that helped me with some of the information in this book. I am very grateful for their help.

Don Meyer Ford and Lincoln located in Greensburg, Indiana is a good place to get service work done, and a great place to buy a new or used vehicle. They allowed me take photographs of their business for Chapter 15, entitled "Finding a Good Repair Shop."

Versailles, Indiana State Police Post and State Trooper, Sargent Paul Adams, who posed for a photograph and gave me some helpful information for Chapter 14, entitled "Driving Habits: Self-Improvement."

I also want to thank the Versailles, Indiana State Police Post for some of the information in Chapter 14.

Wal-Mart Supercenter located in Greensburg, Indiana; they allowed me to take photographs for Chapter 6, entitled "Stop Without Crashing." This chapter talks about brakes.

Most of all, I cannot forget my old friend and buddy, my 2000 Ford Windstar mini-van which made this book possible. My 2000 Ford Windstar is a product of the **Ford Motor Company**. This book will show anyone that it is still possible for their vehicle to run past 100,000 miles with proper care and to help you keep your vehicle longer.

I want to thank my very patient and beautiful wife, Lesa M. Hansen (who is not a mechanic) for helping me with the complete editing of my book.

Contents

Chapter Sixteen

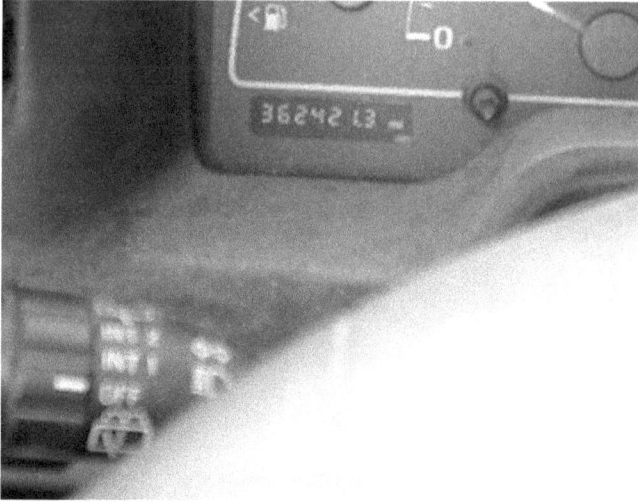

Introduction

It was suggested to me by many people that I should write a book on how to care for a vehicle. The mileage on my Ford Windstar prompted the writing of this book. The information I present in this book is written for the non-mechanics, who do not know much about their vehicles. Non-Mechanics that want to know how to take care of their vehicles, and how to keep their vehicles running longer. It is my wish that the information presented in this book benefits the reader.

Chapter 1

It is All Up to You

It is possible to do vehicle maintenance and to keep your vehicle running longer on a reasonable income and a small budget. Knowing the basics for keeping your vehicle running longer and at a lower cost to you is a good place to start. You can lower the costs of maintaining your vehicle if you are willing or mechanically inclined to do your own maintenance on your own vehicle. If you do not want to take your vehicle to a mechanic's shop all the time for maintenance problems that are easy to fix, then, you must make a decision to learn basic vehicle care and do it regularly and faithfully. This is what I do to reduce my maintenance costs. You are the only one to decide what you are willing to learn, and how much you are going to do on your own vehicle. After reading this book, you <u>will</u> develop a mindset or an attitude that will help you pay more attention to your vehicle. Every time you get around your vehicle you will look at it. (Look at it with just a simple glance and not with binoculars!) You need to pay attention to tires, look for any oil leaks, coolant leaks, blown out light bulbs, and unusual sounds that you hear when your vehicle is running. You depend on your vehicle to take you wherever you need to go. If you take care of your vehicle, it will take care of you.

Since I bought my 2000 Ford Windstar with its new technology at the time, I have had to re-learn some things to be able to take care of my mini-van. There is always something new to learn with today's technology for our vehicles (my old buddy and friend). This is the way I think of my mini-van while I am driving down the road. Do not let this happen to you. (See the photograph above.) It is a photograph of my mini-van on the side of the road when the transaxle and the torque converter failed. (See Chapter 8, entitled "Transmissions and Transaxles: Go, Go, Go").

Chapter 2

Tires: We Start Here

I thought this would be a good place to start because tires are the least understood and cared for on any vehicle today. My Ford Windstar weighs about 4,000 pounds, and it rides on four of these things called tires. What else would you call them? Think about what happens when a tire is blown on your vehicle. A blown tire would cause your vehicle to have steering and handling problems which could cause you to lose control of your vehicle. You do not want to lose control of your vehicle going down the highway at *70* miles-per-hour!

I will give you the basics of what to expect when your vehicle has a blown tire. A front-tire blow-out causes your vehicle to react differently than a rear-tire blow-out. A front-tire blow-out will cause your vehicle's front-end (where the front wheels are located) to vibrate, and your vehicle will start moving in the direction of the blown tire on a straight highway. If you need to make a turn on a highway or rural road, your vehicle will move in the direction of the blown front tire. It will be hard to steer and control your vehicle.

A rear-tire blow-out will cause your vehicle to vibrate at high speeds on a highway. Your vehicle will lose its speed suddenly on a highway or a rural road. The rear end of your vehicle (where the rear wheels are located) will move from left to right. It will feel like you have lost control of your vehicle. When you use the steering wheel, it will have no effect on the rear tires. If you need to make a turn on a highway or a rural road, the rear end of your vehicle will move slowly. You can save yourself from all of this trouble by learning what causes tire blow-outs.

Do you know what causes tire blow-outs? Tire blow-outs are caused by under-inflating or over-inflating a vehicle's tires, cuts in a tire, damaged tire tread, or worn down tires with very little tread remaining on them.

To start learning about tires, we begin by using the correct tire sizes for my mini-van. Cool, huh! On every tire, there will be tire information on the tire sidewalls. You will see the tire's size, air pressure, its weather rating (whether the tire is rated for mud, snow, water, or dry road conditions). This is a standard practice for some of the major tire companies like Goodyear, Goodrich, and Michelin. I have used the tires of these companies on my mini-van. Some tires even have the direction the tire is to be mounted on the tire rim. A few performance tires give you the mounting direction. My son's spare Pontiac Trans-Am tire has this information printed on the tire's sidewall. (See the tire photograph at the start of this chapter 2.)

There are directional arrows that look like triangles on the sidewalls of "high performance tires," and the words "direction of tire rotation" printed on the tires' sidewalls. "High Performance Tires" also have directional tread on them. The tires need to be mounted on the rims in a certain way. The directional arrows help in the mounting of the tires, and the directional arrows will be on the right and left sides of the tires. The tires need to be mounted on the tire rims with the directional arrows facing forward. The arrows always have to be pointing forward (or straight ahead) whether you are driving your vehicle or whether it is parked.

This is additional information about tires. I use all-season radial tires. Goodyear, Goodrich, and Michelin make all-season radial tires. I have used their radial tires on my mini-van. There are specific tires for winter driving conditions. These three major tire companies and others manufacture winter tires. Winter tires are made of a softer rubber and have a different tire tread that grips the road in snowy and icy weather conditions.

Now, let's talk about the tire sizes for my Ford Windstar mini-van. There are different size tires for different makes and models of vehicles. Some vehicles have more than one tire size. I will use my mini-van as a reference to help you understand tire sizes in this chapter. I can use these two types of tires on my Ford Windstar.

1. P205/70R15
2. P215/70R15

The first letter, *P*, stands for *Metric* a type of vehicle tire. My mini-van uses these types of tires. The numbers *205* or *215* stand for the "tire width" which is expressed in millimeters, a Metric Unit of measurement. The tire width is measured from the widest part of the tire (tire tread). Next, the slash separates the tire width from the tire *aspect number* or *tire ratio*. These two numbers are important to know. The tire *aspect number* or *tire ratio* is "in relationship to the tire tread width." The number *70* or *70-percent* is the *sidewall height* (the distance from the rim to the tire tread or *70-percent* of the tire tread width). The capital *R* stands for *Radial* tire. The word *radial* refers to *the type of construction of the tire*. The last number is the tire rim size. The *tire rim size* for my mini-van is *15 inches in diameter*. Whenever you buy tires knowing your vehicle's year, the make, the model, and the tire size for your vehicle is helpful. This information will help you find the exact tires you need for your vehicle.

I want to introduce you to my Ford Windstar mini-van Bible, the Owner's Guide!

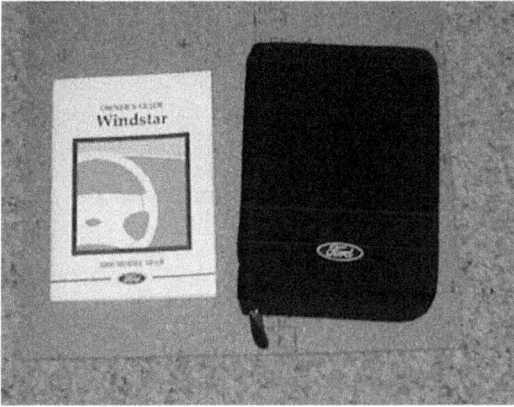

My owner's guide is the basic information I need for my Ford mini-van. It has a wealth (and I do mean a wealth) of information that I have referred to many times during the fourteen years of ownership of my Windstar. Some of the information that I am presenting to you in this book comes from the Owner's Guide. If you do not have an owner's guide for your vehicle, I would contact the manufacturer of your vehicle and find out how to get one. If you cannot find an owner's guide for your mini-van, car, or truck, you can go to this website: Http://www.carmanuals.com. It is a good place to go to find your car, truck, mini-van, and other vehicle manuals. This website has manuals for domestic and foreign vehicles. Do not be afraid to use your owner's manual. The owner's manual was not meant to be used as a flyswatter for killing those killer bugs inside your vehicle while you are driving!

In my owner's guide, it says to *check the tire pressure when the tires are cold*. This means *when your vehicle has been parked for one hour or has been driven less than three miles*. Use a tire gauge and check the tire pressure about two or three times a week. You can find the information about recommended tire pressure limits in your owner's guide or on the door sticker on the driver's side of your vehicle. My mini-van tire information sticker is located on the driver's side, inside the *door pillar*, or the Owner's Guide calls it the *B-pillar*. The door sticker recommends *35 psi for front/rear tires*.

 Thirty-five psi means *35 pounds of air pressure per square inch is needed in the front and rear tires*. That is the maximum tire pressure I need for the tires on my mini-van. (See the photograph below.)

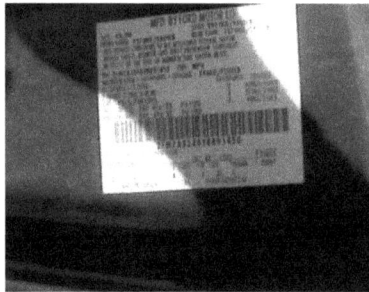

 Tire Gauges are used to check the tire pressure in a tire. This is how you use a Dial Tire Gauge. When you hold the gauge in your hand, you will see the following: One end of the tire gauge has an opening; this is the end that you put on the valve stem of the tire. Press down on the tire gauge that is on the valve stem, and the gauge dial will move. The letters, *psi*, *pounds per square inch*, represent the *American standard of measurement* for measuring air pressure in a tire. You will see this information on the face of the gauge marked *psi*. The second row of numbers is called a *bar*. A *bar* is the *international Metric standard of measurement* for measuring air pressure in a tire and other things.

There are two common tire gauges, the Dial Tire Gauge and the Pen Tire Gauge. The Pen Tire Gauge will be discussed next. The Pen Tire Gauge works similar to the Dial Tire Gauge. There are some differences.

When you press the Pen Tire Gauge down, onto the tire valve stem, a little slide gauge pops out from the bottom of the pen. It shows the actual air pressure of the tire. The air pressure of the tire is measured in the American *psi* standard, the *bar* standard, and *kpa* standard. The *kpa* standard is called *Pascal's* which is another international standard of measurement.

When you do check your vehicle's tires and find that the tires are over their required tire pressure, you can remove excess air pressure from the tires by using the Pen Tire Gauge. The *deflator*

is on the opposite end of the Pen Tire Gauge. It is on the opposite end that you use to check the air pressure of the tires. Press the *deflator* against the tire valve stem *to remove excess air pressure.* Then, check the tires again with the air pressure side of the gauge (that fits onto the valve stem) to get an air pressure reading. Keep checking and deflating your vehicle's tires until the tires have the correct air pressure needed for your vehicle.

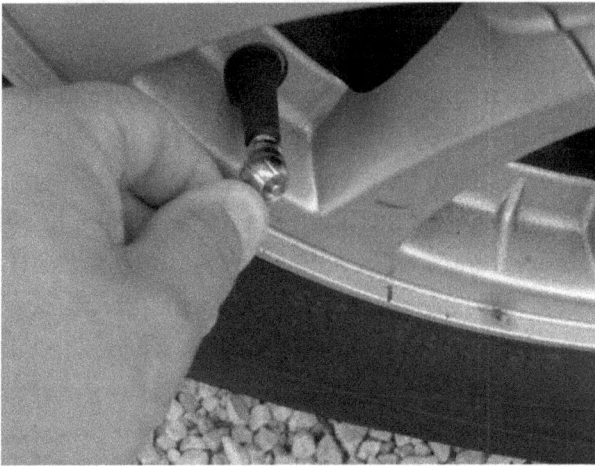

The Dial Tire Gauge does not have a *deflator*. I use both gauges. The Dial Tire Gauge is good for a quick and fast air pressure reading of your vehicle's tires. The Dial Tire Gauge uses the American *psi* standard and the *bar* standard. The Pen Tire Gauge is more versatile. (See the photograph below of the *deflator* side of the Pen Tire Gauge.)

Now, let's learn about the importance of tire tread on your vehicle's tires. You need to look at the tire tread to determine if there are problems with your vehicle's tires. How do you know if you have a problem with the tire tread on your vehicle's tires? Does the tire tread look worn on the edges of the tires? If the answer is "yes," you are driving on tires with low air pressure in them. Another way to tell if the tire tread is worn is to take a look at the center of the tire tread. If the tire tread is worn in the center, this means you are driving on a tire with high air pressure in it. (It is over inflated.) High air pressure and low air pressure tires are not safe tires to be driving on!

Another thing to learn about tire care is *tire tread depth.* I want to explain the basics on *tire tread depth,* or it is also called looking for *wear bars. Wear Bars* are indicators that the tire manufacturers put on a tire *to show that the tire is worn and needs to be replaced.* To find *wear bars* on your vehicle's tires, look at the tires (tire tread area) and look at the spaces that are between the tire tread. You will see small thin lines that are raised up running from tread to tread (going in the direction of left to right) these are w*ear bars.*

Did you find them yet? The subject of *wear bars* is a dead giveaway. What do I mean? Looking for tire-wear bars on a tire is something you can see very easily. When tires are worn, the *wear bars* are seen more easily. *Wear Bars* are like a sore thumb on a worn tire! When you see the *wear bars* on the tires of your vehicle, driving can be very dangerous; especially in wet, rainy weather, or icy, snowy weather conditions. I think of *wear bars* as trying to run around in a house in your socks and trying to stop on over waxed hardwood floors. When you need to stop, you cannot; because you have no traction. You run into walls that do not move and are not flexible. I do not recommend this teaching moment for kids. Adults can try sliding in their stocking feet in their homes to see how worn tire bars perform at high speeds. Do not count on your vehicle's brakes to help you stop your vehicle when you have worn tire bars on your vehicle's tires! The *wear bars* in the photograph below go across the entire tire tread. They are marked in white so that they can be easily seen. The photograph shows you the *wear bars* on my vehicle's tires.

A very simple way to check for tire tread wear is with a coin, specifically a penny. Put the penny on its edge between the slanted tire tread in the center of the tire so that Lincoln's head is upside down. If Lincoln's head is above the tire tread, then, the tire is worn. I am giving you the very basics details in tire care. You have to remember to do "tire checks" faithfully. These details can and will save your vehicle's tires and your life.

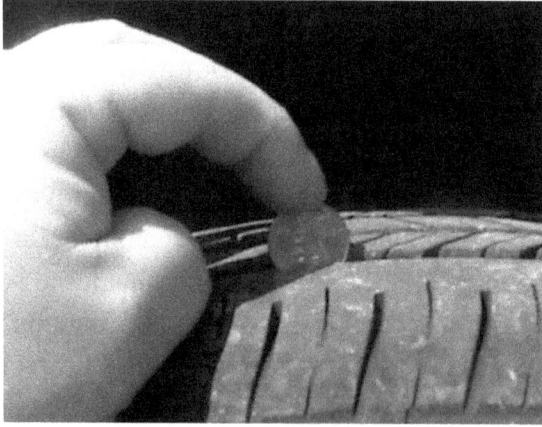

Now, think about the importance of this tire information, and how easy tire care is to do on your vehicle. I hope these photographs help your understanding and explain what I am teaching you.

Here are brief lessons on spare tires and tire rotation. Today, most vehicles have a compact spare tire. Spare Tires are **only** meant to be used in emergencies. They are **not** to be used for long distance driving! Spare Tires are designed to get you to the closest gas station or repair shop. The recommended speed limit for a spare tire is *50 miles-per-hour*. They are smaller than everyday vehicle driving tires, and they are mounted on smaller rims. The photograph above shows a spare tire on its smaller rim. Spare Tires have higher air pressure in them. You feel more of the bumps in the pavement. There is not much tread on a spare tire. The unevenness of the pavement on the road and other road hazards can cause the spare tire to blow-out.

Rotating the tires on your vehicle can extend the life of the tires, and lower the costs of maintaining the tires on your vehicle. There are two ways to rotate tires. The first method is called the *Front to Back Method*. This means that *the front tires are mounted on the back axles* (or where the back tires were). *The back tires are then mounted on the front axles* (or where the front tires were).

The second method for rotating tires is called the *Crisscross Method or X Method*. This means that *the tire in the front on the driver's side of the vehicle is mounted on the back axle of the passenger's side of the vehicle*, and then, *the back passenger's side tire is mounted on the front axle of the driver's side of the vehicle*. This is the first part of the *X*. The second part of the *X* is made when *the front passenger's side tire is mounted on the back axle of the driver's side of the vehicle*, and then, *the tire on the back axle on the driver's side of the vehicle is mounted on the front axle of the passenger's side of the vehicle*. You can pay a tire store, a department store with an automotive garage department, an automotive garage, or a dealership to rotate the tires for you if you do not have the equipment to rotate the tires on your vehicle yourself. The owner's manual will normally tell you how often to rotate the tires on your vehicle. If your owner's manual does not give you any tire rotation information, the general rule is rotate the tires on a vehicle every 3,000 to 5,000 miles or every time you have the oil changed in your vehicle.

For the do-it-yourself mechanics, you will have to keep a record of the miles you drive and the oil changes you do. This will help you to decide when to rotate the tires on your vehicle.

Chapter 3

Oil: The Lifeblood of an Engine

Oil is the lifeblood of an engine. Without oil in an engine, it would fail in seconds. (You do not need a stopwatch.) Buying oil is cheaper than replacing an engine; (The famous last words!) In my owner's manual, it says that my mini-van needs 5 quarts of SAE 5W-30 oil to completely fill my 3.8 liter V6 engine. I have to remember that five quarts is needed for my mini-van's engine before it is completely bone dry, without oil. You can see that having an owner's manual is very helpful. Open your owner's manual if you do not know how much oil your vehicle's engine needs. SAE stands for **S**ociety of **A**utomotive **E**ngineers and the numbers 5W-30 is a multi-viscosity grade of oil. The letter *W* stands *for winter or for colder temperatures*. The number *5* represents that *the oil will act like a grade 5 automotive oil in colder temperatures.* Generally, *when an engine is at 180 to 210 degrees* (its normal operating temperature) *it will act like a grade 30 automotive oil*. A lower grade of oil is needed for colder temperatures to prevent excessive engine wear. A lower grade of oil is needed in winter or colder temperatures to help a vehicle's engine on start up because there is no oil on the engine parts. In winter or colder climates, a vehicle's engine does not reach its operating temperature quickly. A thinner grade of oil will help your vehicle operate more efficiently; therefore, the oil will get to the engine parts faster.

Do you know when to check the engine's oil level? The best times to check the engine's oil level is before you start your vehicle or at least 45 minutes after you turn "off" the engine. I keep thinking about the guys at gas stations, who check the oil level of their vehicles, while they are pumping gas; they do not know what they are doing. It normally takes 45 minutes for all of the oil inside of an engine to drain into the crankcase pan; otherwise, you will get an inaccurate reading on the oil dipstick. Do you understand what I mean? Locate the oil dipstick on the engine of your vehicle by reading your owner's guide or by looking for it on the engine. It should be close to the engine block in the front or on the side of the engine. The photograph below shows a "pull handle" of a *dipstick*.

18

Remove the *dipstick* and wipe it clean with a clean rag. Look at the *dipstick*—notice the words and *hatch marks* on it. My *dipstick* has *hatch marks*. Does your *dipstick* have *hatch marks* and words? For example, the abbreviations *min.* and *max.* are used for the words *minimum* and *maximum*. The abbreviations represent the minimum oil level and maximum oil level in the crankcase pan of the engine. If the engine's oil level is below the minimum oil level and you have driven your vehicle for a while on a low oil level, perform the last rites on your vehicle's engine; because you might have damaged the engine forever. You will need a new engine or a new vehicle. Make sure your vehicle is on level ground before doing an "oil check" on your vehicle, or you will get an inaccurate oil reading. Generally, between the abbreviations min. (minimum) and max. (maximum) levels on the *dipstick* is the difference of one quart of oil in an engine; unless, it is specified on the dipstick or in the owner's manual. This means that most of the time one quart of oil is needed if the *dipstick* shows that the engine's oil level is in the minimum area on the *dipstick*. To bring the engine's oil level to the maximum level, one quart of oil is usually needed. It seems like we can't get away from the owner's manual. Can we? Remember, oil in the hatched (diamond) area of the *dipstick* is considered the normal level for the oil in a vehicle's engine. If there is oil in the *hatch mark* area of the *dipstick* you do not have to add oil to the engine.

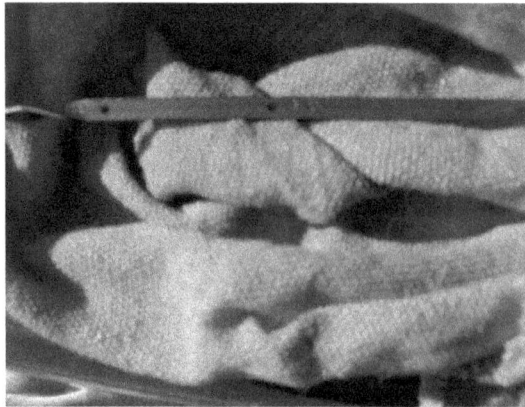

These photographs are good pictures of my mini-van's oil dipstick. Use this picture as a reference to help you understand what I am trying to teach you about dipsticks. This is a teachable moment that you will remember for the rest of your life: *the hatching and the two holes* (that look like black dots in the pictures) *between the two words minimum and maximum* represent my mini-van's *oil level. It is within the normal oil level range.*

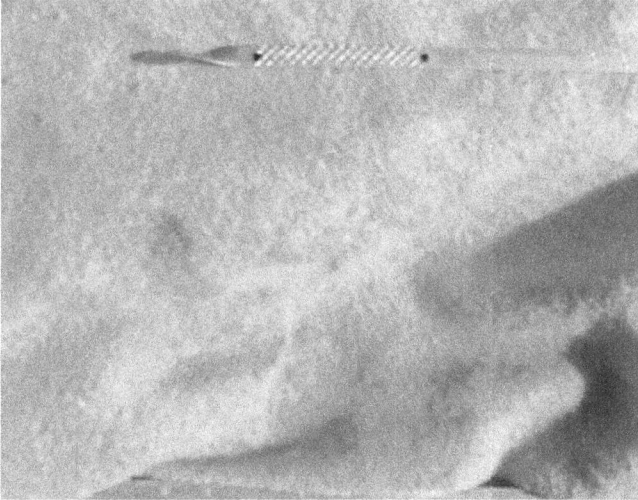

Reinsert the oil dipstick after you have wiped it clean. Pull the *dipstick* out again from its holder on the engine. If you discover that the engine is low on oil, pay attention to where the oil level is on the *dipstick* when you reinsert it back into engine. The closer the oil level is to the minimum oil level mark means you need to add one full quart of oil to bring the engine's oil level up to the maximum oil level or within the normal oil level range.

If there is no oil on the *dipstick*, your vehicle's engine needs more than one quart of oil. When you hear a metallic noise (metal hitting metal), this is the *valve train* of the engine. The *valve train* will sound like a metallic knocking noise. The *valve train* is the first place an engine will be starving for oil. The *valve train* *consists of lifters, push rods, rocker arms, and valves*. It will not operate properly or stop operating because of the lack of oil in your vehicle's engine. That means the engine in your vehicle stops operating!

To add oil to your vehicle's engine, simply locate the oil filler cap. The oil filler cap is located on the valve cover of your vehicle's engine. Use your owner's manual to help you find the oil filler cap. On newer vehicles, the oil filler cap will have the words *engine oil* marked on it. My Ford Windstar has the words *engine oil* marked on its filler cap.

You simply unscrew the oil filler cap. On most vehicles, the filler cap comes off by turning it counter-clockwise. If it does not come off when you turn it counter-clockwise, then, turn the cap clockwise. Remove the cap and clean the seal that is on the oil filler cap. Cleaning the seal prevents oil from leaking from the oil filler cap. **Make sure your vehicle's engine is not running. Do not pour oil in your vehicle while the engine is running, or clean the oil filler cap while the engine is running. The hot engine could ignite the spilled oil and could cause a fire on the engine.** The oil filler cap seal stops oil from coming out of your vehicle's engine when you are driving.

Would you like to learn the correct method for pouring oil out of an oil bottle? The pour spout of the oil bottle has to be on top. (See the photograph above.) If the oil bottle has an offset pour spout, the pour spout is not in the center of the oil bottle but at the top of the bottle. This makes the oil easier to pour because air gets into the bottle and helps the oil flow more evenly. Today, most oil bottles are made with an offset pour spout. Some oil bottles still have a center pour spout. To pour oil from a center pour spout, you have to angle the bottle at an upward slant and pour very slowly to allow air to get into the bottle to help the oil to flow more evenly.

If you are good at pouring, like me, you can pour the oil into the engine without spilling oil everywhere and making a mess on everything. You can also use a funnel. Get a plastic funnel at a Wal-Mart store or an auto parts store. Paper funnels are available at gas stations to help you pour oil into the engine without making a mess. They are usually free.

Generally, on oil bottles you will find a clear spot on the side of the bottle, this is a gauge. The gauge shows you how much oil is inside the bottle. The gauge can be handy and helpful when you are pouring oil into the engine. You can use the sight gauge, on the side of the bottle, to show you where to stop pouring if you need less than a quart of oil to bring the engine oil level within the normal oil level range. After pouring oil into the engine, you need to check the oil level with the *dipstick*. (Success! You did it yourself, and you did not have to pay anyone to do it for you.) Now, you can put the oil filler cap back on. Do you remember when I talked about the seal on the oil filler cap?

If you look closely at the photograph, you will see a rubber seal near my fingertips. I clean this seal with a clean rag. (Why would you use a dirty rag?) I clean the valve cover too. The oil filler cap needs a clean surface to make a tight seal. The tight seal on the filler cap prevents any major problems or a fire on the engine. Do not forget to wipe off any spilled oil on the valve cover. Do not overfill an engine with oil. Overfilling an engine with oil causes an engine to build up excessive pressure, and an engine might blow a gasket or other engine seals. It will result in a major oil leak (all of it!) and extensive damage or failure of an engine.

Now, you know why checking an engine's oil level two or three times a week is very important to extend the life of your vehicle. If you do a lot of long distance driving, you will have to check the oil level of your vehicle's engine more than two or three times a week. It is a good idea to keep an eye on the oil level in the engine while looking at other things under the hood of your vehicle that will also help to extend the life of your vehicle.

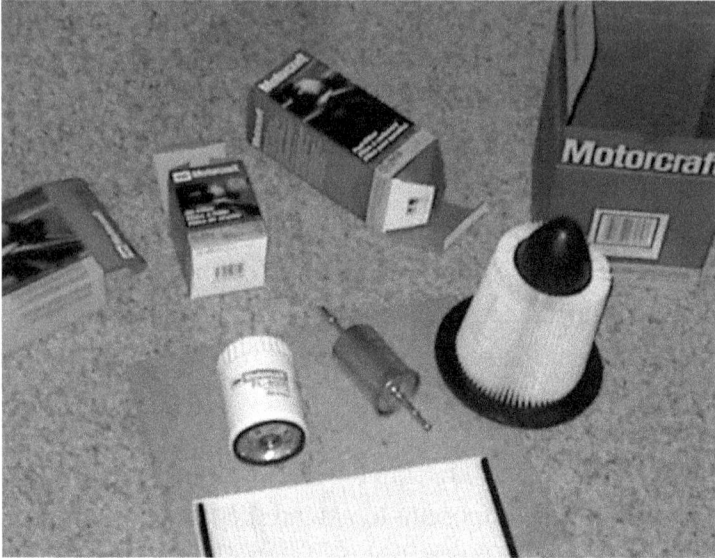

Chapter 4

Filters: Nice to Have Clean Air

This chapter is pretty straight forward and simple about filters. Filters are necessary for the maintenance of your vehicle. For example, the main reason for an air filter is to help prevent dirt and bugs from entering the engine. The most common filters you will see in vehicles are air, oil, and fuel filters. On my Windstar, I have an additional filter, a cabin air filter; I still cannot believe it! The cabin air filter is supposed to filter the air before the air goes into the interior of my mini-van. You probably know by now, how to find the type of filters your vehicle definitely needs. That's right, by looking in your owner's manual. Remember, the owner's manual is a good place to start if you do not know that already. Ordinarily, filters are easy to change once you have located them on your vehicle. The owner's manual can show you all of the filters' locations in your vehicle, and how to change the filters too.

The reason I use Ford parts is I know they will fit on my Windstar. The parts are built to Ford's specifications, and I can get new parts like the parts that were originally installed on my Windstar. Some parts on your vehicle can be replaced with rebuilt parts to save costs. I do not go to the automotive discount stores to buy major parts for my Windstar. I have been "burned" in the past. What do I mean by being "burned?" I have bought parts that were supposed to fit on my mini-van, but when I tried to put them on, the parts **would not** fit. The automotive discount stores' generic no-name-brands were not money saving deals to me! No-name-brand filters are a waste of money and time. The poor quality of the filters will have you replacing the filters in your vehicle more often than you should. You will not be saving money. You will be spending more money. Do you know the best time to replace the filters in your vehicle? I usually replace all of the filters about twice-a-year. Believe it or not, filters are not that expensive. Where you live will affect how often the filters are changed in your vehicle. You will need to change the filters in your vehicle more often if you live in an area that is a desert and has sand storms or on a sandy beach with blowing winds. People that live in dust bowl areas will have to change their vehicle's filters more than twice-a-year too. This is a simple rule to follow. Filters can be replaced twice-a-year or as recommend by the vehicle's manufacturer in your owner's manual. The filters help your vehicle's engine operate more smoothly and efficiently.

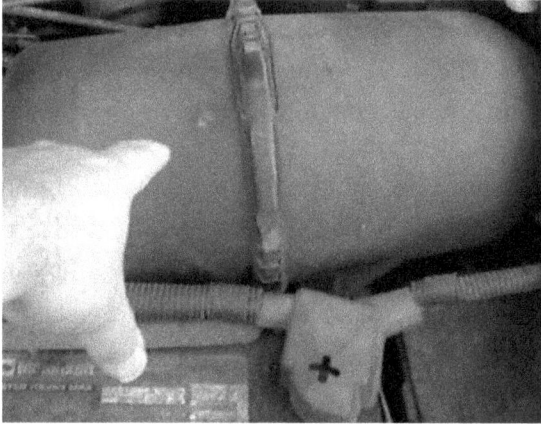

Normally replacing filters is easy to do. For those of us who are on the run, you can replace your vehicle's filters in a driveway or in a parking lot. This photograph shows the housing for the air filter on my Windstar. Ford's Windstar mini-van air filter housing has only one latch to release the air filter housing. I can remove and replace the air filter very easily. On other makes and models of vehicles, there will be either latches or snap-fittings to hold the housing together for the air filter. I am telling you basic and simple information because you may not own a Ford Windstar mini-van. The photographs are for you to use as a reference. Since you have the air filter housing open on your vehicle, this is a good time to clean it. A clean air filter housing helps the new filter to stay cleaner longer. Every time you change your air filter clean the air filter housing. The second filter I want to talk about is the oil filter.

Since my vehicle's oil filter is on the backside of the engine, I have to get to the oil filter by going underneath the vehicle. This means I will have to crawl under my vehicle. You will have to do the same thing if the oil filter of your vehicle is on the backside of its engine. You can use a *mechanic's creeper which is like a skateboard for your body*. You can use the *mechanic's creeper* while working underneath your vehicle or inspecting parts under your vehicle. The *mechanic's creeper* works best on a hard surface. It does not work well on a gravel driveway. The easiest way to look for the oil filter or to change the oil in your vehicle is to drive your vehicle up onto a pair of *vehicle ramps*. You can get this equipment at automotive discount stores. This information is for the ambitious do-it-yourself person, who wants to crawl underneath his vehicle to change its oil filter.

Not all vehicles have oil filters that require you to crawl underneath the vehicle. Some oil filters are located on the front of the engine. This makes it easy for you to lift the hood of your vehicle and change the oil filter. Whether the oil filter is located on the backside (underneath your vehicle) or the front (under the hood) of your vehicle's engine does not matter. You will need a container to drain the used oil into as you empty the engine of old oil. An *oil drip pan* is what you will need. You can get one at an automotive discount store. After draining the oil from the engine, you must take the old oil to an oil recycling center. **Do not put the used engine oil in storm drains, landfills, or other places that will affect the health and well-being of the public. Do not forget to change the oil filter when you change the oil in your vehicle's engine.** If you do not want to crawl underneath your vehicle, you can get your engine oil and oil filter changed at an automotive service center that changes the oil and other fluids in your vehicle. The automotive service centers will also dispose of the old engine oil and the old oil filter for you. Engine oil and oil filters are something you change every 3,000 to 5,000 miles. Check your owner's manual for more details.

   Take your time changing the oil filter! **If you do not put your vehicle's oil filter on tight enough, you will lose engine oil around the filter, and you could lose your vehicle's oil filter too. This will lead to an oil trail to the engine's grave! Oil is the lifeblood of your vehicle's engine.** Normally on an oil filter, you will find a black seal on the top of the oil filter. (See the photograph below.) This seal prevents oil loss when the oil filter is on the engine doing its job of filtering the engine oil. The oil filter filters out metal shavings, carbon, dirt, and loose rust in the oil from your vehicle's engine. Remember to tighten an oil filter first with your hand before using tools. Turn the oil filter with your hand until it stops spinning, and then, turn the oil filter roughly about a quarter of a turn (at the most) with an oil filer wrench or a strap wrench to snuggly fit the oil filter seal against the engine.

This is a photograph (previous page) of the black seal on an oil filter for my Windstar. Changing the oil and the oil filter as recommended by the manufacturer is essential to maintaining your vehicle. If you do not want to change the oil and oil filter on your own, this will give you an idea of what an automotive service center does to your vehicle. Some vehicle dealerships can change your oil, oil filter, air filter, and fuel filter all at the same time.

The third filter I want to talk about is the cabin air filter. Not all vehicles have a cabin air filter. The cabin air filter is easy to change. I remove the tray near the windshield in the engine compartment and take out the old filter and place the new one in the tray. Do not forget to close the tray. I wish everything else on my Windstar was that easy to work on.

The last filter to look for on your vehicle is the fuel filter. The fuel filter removes the junk that is floating in the gas while it is going to the engine form your vehicle's gas tank. The location of the fuel filter can vary. The fuel filter can be in three different locations. It can be in the engine compartment. It can be near the gas tank, or under the vehicle. It depends on the make and model of your vehicle. The fuel filter for my Windstar is underneath the vehicle. When you put in a new fuel filter, be sure the flow direction arrow is pointing towards the engine. The arrow indicates the direction of the fuel flow through the filter. (See the photograph below.) If you do not put the fuel filter in correctly, you will not have any fuel flowing to your vehicle's engine. You will not go anywhere! Putting a fuel filter in backwards **will** cause gas to stop flowing or cause low gas flow to your vehicle's engine. The junk that gets in your vehicle's gas tank comes from the underground gas tanks at the gas stations that store the fuel for your vehicle. The fuel filter helps get rid of the junk in the fuel. This pretty much explains the basics of filters. This is a simple and no nonsense way of dealing with filters. Changing air filters, oil filters, and fuel filters **will** help increase the gas mileage of your vehicle. There is still more to learn about caring for your vehicle on a budget.

Chapter 5

Engine Coolant: Keep the Engine Cool

I am going to spend some time talking about engine coolant basics. A subject that is important and vial to the engine of your vehicle. The cooling system is just as important as the oil system in a vehicle. How important is coolant to an engine? Most people never think about the operation of an engine from the inside of an engine, and what happens under the hood to keep an engine cool. Imagine it is 115-degrees Fahrenheit outside, and you are sitting in your vehicle at a stop light with your vehicle in drive and your foot on the brake waiting for the light to change. The air conditioner is on maximum capacity to keep you cool in the sweltering heat. All of this activity is a load on your vehicle's engine. Your vehicle needs coolant to cool the engine down while it is operating. Without coolant in an engine, the engine would run very hot and eventually fail (or seize up). People never stop to think about what is actually needed to make their vehicles operate until something breaks or fails on their vehicle, and they come to a complete stop somewhere. Today's engines are what I call a *composite design*. I do not know if anyone else calls engines *composite designs*. What do I mean when I say that an engine is a *composite design*? "The engine is made of different metals and plastics." The main block (where the pistons are located) could be made of cast iron. The cylinder heads could be made of aluminum. The valve covers and intake manifold could be made from plastic. Nice design, huh! Whenever you open the hood of your vehicle, take a look at the different colors of metal and plastic used to build an engine. (Who knows what else today's engines could be made of?) Surprise, surprise! My Ford Windstar mini-van is a composite engine design. This might be a cheaper way to build an engine, but I think for an engine's longevity, it should be made out of cast iron. The main block and cylinder heads should be cast iron. The intake manifold should be cast steel, and the valve covers should be made of sheet metal.

These types of metals are far more forgiving and tolerant to an engine working on low engine coolant. What do I mean by *forgiving and tolerant metals*? *Forgiving and tolerant metals* mean that *the metals will not warp or become distorted at high temperatures because of a low coolant level in the engine*. Earlier engine designs before the 1990's were made of cast iron, cast steel, and sheet metal. The engines were more durable.

I am familiar with cast iron main blocks and cylinder heads, and also cast steel intake manifolds and valve covers made of sheet metal in older engines. I have worked on earlier vehicle engines with these durable metal parts. Today, some engine parts are built using cast aluminum or high temperature plastic. The engine parts that are made from cast aluminum and high temperature plastic are about 30-percent cheaper to manufacture, but they are less forgiving and tolerant. If you do not take care of your vehicle's engine you will be replacing parts more often. In today's newer engines, **you must keep the engine's coolant system working correctly;** or you will have an **absolute nightmare** on your hands.

When an engine is working with a low coolant level or with dirty coolant in its system, you run the risk of warping the aluminum heads. (These are the bright, shiny, things on the top of your vehicle's engine.) Engine Coolant made by Prestone is what I use in my mini-van. It is green in color because of the chemical Ethylene Glycol in the coolant. Over time as an engine is working and doing its job, the coolant does not look green in color (or have a green tint) any more. The coolant looks like the color of rust (or rust tint). The engine coolant needs to be changed when it looks rust in color. Today's newer vehicles that come from the manufacturers use an orange tinted color coolant. The chemical Dexcool in the coolant makes the coolant orange in color. The orange tinted coolant will need to be changed when it looks brown in color. The cast iron main block develops rust inside of it because of water and coolant in the main block. No coolant or very little coolant will warp the cylinder heads. Warped cylinder heads are a clue that your vehicle needs a new engine or a rebuilt engine. Ordinarily, rebuilt engines cost less than a brand new engine. Depending on the year, the make, and the model of your vehicle, some rebuilt engines might not be available. Changing the engine coolant is even less costly. You will save thousands of dollars. That is why I am taking the time to explain how important coolant is to an engine. I want to prevent you from having these problems.

Coolant is made from water and an anti-freeze liquid. The basic parts for a cooling system in a vehicle are the radiator, water pump, thermostat, hoses, and an overflow reservoir (a place to add coolant for the engine's cooling system). The cooling operation of a vehicle is as follows: as the engine is working, the water pump is running, pumping coolant from the radiator through the engine block and cylinder heads and through the thermostat. (The thermostat opens and closes like a door to keep a steady temperature in the engine.) The coolant goes back into the engine's radiator to start the process over again. A cooling system is supposed to be a *close loop system*. This means it should "not lose coolant," but it does over time. That is why you have to check the engine's cooling system at least once-a-week. You can do this by checking the level of coolant in the engine.

This is how you check the amount of coolant needed in an engine's cooling system by looking at and checking the overflow reservoir in the engine compartment. There are two marks labeled *cold and hot levels* on the overflow reservoir. (See the photograph on the next page.)

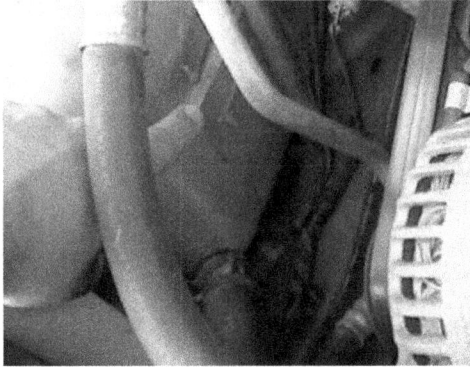

I am pointing to the area of the *cold and hot level* marks on the overflow reservoir for my Ford Windstar mini-van. The coolant level marks are white on the overflow reservoir. Most vehicle engines have the *cold and hot levels* marked on their coolant overflow reservoirs. It depends on the make and model of the vehicle. The marks labeled *cold and hot levels* "determine if an engine needs to have coolant added to its system." The *cold level* mark on the reservoir means that "when the engine has not been running, the coolant level should be at the *cold level* mark." If the coolant level is below the *cold level* mark, while the engine has not been running; you need to add coolant until the level of coolant is at *cold level* mark. The *hot level* mark on the reservoir means that "the engine has been running, and then, turned off." The engine has not cooled down yet. If the coolant level is at the *hot level* mark, do not add any coolant to the overflow reservoir. If the coolant level is below the *hot level* mark, add coolant until it reaches the *hot level* mark. Put coolant in the overflow reservoir only to the *cold level* mark or the *hot level* mark. Do not overfill the cooling system reservoir. The overflow reservoir's purpose is to allow the excess coolant to go into the reservoir to prevent damage to the cooling system. The excessive coolant could cause an engine's hoses to break (or blow) and the loss of engine coolant all over the highway. You will be watching a green or orange trail of coolant from under your vehicle, until your vehicle comes to a complete stop on a highway or somewhere on a less traveled or deserted road. You will have a well-cooked (overheated) engine.

What kind of coolant should you use when your vehicle needs coolant? A good mixture of coolant to use is a *50/50 mixture* for general purpose use for warmer and moderate climates.

A *50/50 mixture* is *50-percent coolant and 50-percent water*. You can buy the coolant premixed, and just add it to the cooling system in your vehicle. Remember, the coolant mixture goes into the overflow reservoir. I use Prestone engine coolant because it is a good quality product. If you have a vehicle that uses orange tinted coolant, you **must** use orange tinted coolant, and **never** mix Ethylene Glycol (green tinted coolant) with Dexcool coolant (orange tinted coolant). The result is the two chemicals will react with each other and create a gel inside of your vehicle's cooling system. This will cause more problems for you and your vehicle. Poor quality coolants will have to be changed more often.

For those of you, who do not want to use or cannot use a *50/50 mixture* of coolant, there are concentrated forms of Dexcool and Ethylene Glycol. The concentrated forms of these two coolants are used in colder climates. The coolants can be used in temperatures less than minus thirty-four degrees Fahrenheit or minus thirty- six degrees Celsius. You will have to mix a higher percentage of concentrated coolant with water. For example, *a mixture of 60- percent coolant and 40-percent water is recommended for some colder climates.* Check your owner's manual for more information.

For anyone who still has trouble understanding the cooling system in their vehicle's engine, let me explain it in a different way. To see this engine cooling operation in action, you need to open your vehicle's hood and start your vehicle's engine. Watch the engine working while the engine is running at an idle. ("Idle" means "an engine running without moving.") You will see the fan belt (or serpentine belt) moving. **Do not stick your fingers or other objects in the fan belt!** You will end up in a hospital getting your fingers reattached or permanently lose them. If you stick objects in the fan belt, you could lose your life. When the fab belt is working, the water pump is working. They work together to send coolant through your vehicle's engine. The fan belt and the water pump are working, to get the engine coolant pumped from the radiator to the engine block by way of cooling jackets inside the engine block. The cooling jackets are around an engine's pistons and cylinder heads that exist in the engine block. The coolant goes through the thermostat (that opens and closes like a door) and returns to the radiator. This process is repeated over and over to keep your vehicle's engine cool when it is working. I hope this helps those who have a little trouble understanding what I am trying to explain.

The cooling system is not a perfect *close loop system*; over time the coolant evaporates through the overflow reservoir vent cap, worn coolant system parts (like the water pump and radiator) and leakage of coolant through worn and weak hoses.

Another thing you can do while you are looking at your vehicle's cooling system is pinch the engine's cooling system hoses. (I hope you are not doing this while the engine is running at an idle! Ouch, hot stuff in there!) In the photograph below, my thumb is on a large coolant return line to the radiator. My Ford Windstar is a front wheel drive vehicle. It is hard to see the coolant return line to the radiator on a front wheel drive vehicle. By pinching or squeezing and putting force on the cooling system hoses, you are finding out if there are any cracks or weak spots in the cooling system lines.

The cracks show you where future coolant leaks will be and where the hoses will break.

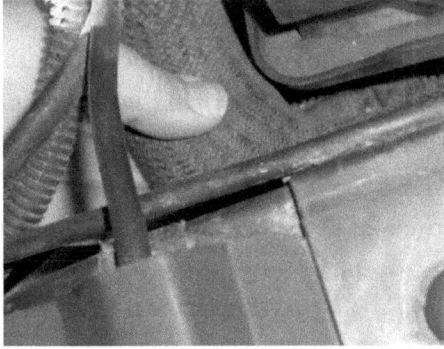

Weak spots in the cooling system hoses will also cause the loss of engine coolant. To find a weak spot in a hose, you need to pinch the hose. If the hose is very easy to pinch and collapses very easily when you pinch it, the hose needs to be replaced. Cooling System Hoses are made of semi-stiff rubber with very little flexibility. When you squeeze the hose, there should be some resistance in the hose. If there is no resistance in the hose; this is a weak spot (or soft spot) in the hose. A weak spot (or soft spot) is the first place a hose will break. Replace a soft hose as soon as possible.

It is better to find out now, while you are checking hoses; before the hoses blow, while you are driving to who knows where and lose your vehicle's coolant. You do not want your vehicle to lose coolant and get stuck in an area you do not want to be in.

Coolant leaks are easy to see; look for a green or orange spot on the ground underneath your vehicle. This is a good indicator that your vehicle has a coolant leak. Another place to check for a coolant leak is around the cooling system hoses. You need to lift the hood of your vehicle once-a-month to check for coolant leaks around the hoses.

To find the thermostat on the engine is very easy, locate the *top hose* (or coolant return line) on your vehicle's radiator leading back to the engine. *The coolant return line is the hose that carries coolant from the main block of the engine to the radiator.* The *top hose* is located where the hose and the engine meet. On most vehicles, the thermostat is located in the hose housing. The thermostat is on the coolant return line or *(top hose)* that is attached to the engine, and this hose leads back to the radiator.

This is what I do. When I need to replace the water pump on my Windstar, I also replace the thermostat. This will prevent future problems with the engine's cooling system in your vehicle.

If you see rust or brown tinted color coolant that means it is time to change the engine's coolant. You can also save time and money when you check the equipment that operates the cooling system in your vehicle once-a-month. Practice good habits in maintaining your vehicle and you will definitely save thousands of dollars and reduce your headaches.

Chapter 6

Brakes: Stop Without Crashing

I want to give you a better understanding of the basics of the braking system on your vehicle. Very few people think about the brakes on their vehicle, until they need to use the brakes! Brakes are a device that stops the tires from moving on your vehicle when you need to stop. Yes, stop your vehicle when you want to; not when you run your vehicle into something to stop! I will try to explain to you how today's *power-assisted braking system* (or the common term is "power brake system") operates when you apply pressure to the brake pedal. The brake pedal is connected to the *master brake reservoir. A master brake reservoir is a device with a power-brake booster which is vacuum powered by a vehicle's engine.* The vacuum increases the flow of the brake fluid in the *power-assisted braking system.* The brake fluid lines are attached to the *master brake reservoir.* This gives you power to slow down or stop your vehicle. The brake fluid is under pressure going through the brake lines which causes the *brake pads* (on the front brakes) to come in contact with the *disc rotors* on the front brakes. At the same time, the *brake shoes* (on the back brakes) come in contact with the *brake drum*s that are connected to the *drum brakes* to help stop or slow down a vehicle.

My Ford Windstar uses a power-assisted braking system; it has *disc rotor brakes* on the front wheels and *drum brakes* on the back wheels. My Ford Windstar has in its braking system an anti-lock braking system. It is abbreviated *ABS* or *Anti-Lock Braking System.* What is an anti-lock braking system? It means that in slippery weather conditions or slick road conditions the *anti-lock braking system prevents full stoppage of the front and back wheels of a vehicle to help maintain control of the vehicle in these hazardous conditions.* The *anti-lock braking system* prevents skidding and losing control of your vehicle while you are using your vehicle's brakes to stop.

In slippery weather conditions or slick road conditions, pump your vehicle's brakes to help you stop your vehicle. This is how you pump your vehicle's brakes. Press your foot on the brake pedal and then, release the brake pedal quickly; keep repeating the pumping action while you are going through the hazardous conditions. Remember to keep the steering wheel straight.

If you are in a skidding situation because of hazardous conditions, use the steering wheel and steer your vehicle in the direction of the skid. Look at the back end of your vehicle. The skidding direction of the back wheels (either to the left or to the right) will determine the direction you turn the steering wheel. If your vehicle's back wheels are turning to the right during the skid, turn the steering wheel to the right. If your vehicle's back wheels are turning to the left during the skid, turn the steering wheel to the left. Turn the steering wheel while you are pumping the brakes quickly and repeatedly until your vehicle comes to a complete stop.

The *anti-lock braking system* has been around since the 1970's, and it is reliable and trouble free if you take care of it. Now, I would not buy a vehicle without an anti-lock braking system. Up to seventy percent of the braking power on most vehicles is on the front brakes to slow down or stop the vehicle. The photograph at the beginning of this chapter is a picture of a front disc rotor brake on my Ford Windstar mini-van. Today, many vehicles have disc rotor brakes in the front and drum brakes in the back on the wheels of the braking system. Some vehicles have disc rotor brakes on the front and the back wheels of the braking system. It depends on the design and year of your vehicle. At some point in the year 2000, Ford decided to put disc rotor brakes on the front and back wheels of its Ford Windstar mini-vans. My vehicle has disc rotor brakes on the front wheels and drum brakes on the back wheels of its braking system.

The following is a general rule: Today, a full-size truck will have disc rotor brakes on all four wheels of its braking system. A small truck will usually have disc rotor brakes on the front wheels and drum brakes on the back wheels of its braking system. A large car will have disc rotor brakes on the front and back wheels of its braking system. A small car will have disc rotor brakes on the front wheels and drum brakes on the back wheels of its braking system. A mid-size vehicle built today will have disc rotor brakes on the front and back wheels of its braking system. Earlier models of mid-size vehicles and large trucks have disc rotor brakes on the front wheels and drum brakes on the back wheels of their braking systems. A performance vehicle built after 1990 has disc rotor brakes on the front and back wheels of its braking system. Before 1990, a performance vehicle had disc rotor brakes on the front wheels and drum brakes on the back wheels of its braking system.

In this chapter, you will learn what to look for before your vehicle has braking problems. For the ambitious, do-it-yourself mechanic, the *Chilton's Repair Manuals* are what you will need. The *Chilton's Repair Manuals* are written for specific vehicles. The manuals will stop this movie scenario in your front yard or driveway. The do-it-yourself mechanic who yanks out a pile of wrenches and takes everything apart in his driveway or front yard and then, wonders why he has spare parts leftover after putting his vehicle's brakes back together again. A person I was riding with told me what he had done to the brakes on his vehicle. He said that he had leftover parts from trying to replace his vehicle's brakes. I stopped riding with this individual. *The Chilton's Repair Manuals* teach you how to fix many vehicle problems including brake problems. You can go to this website:

Http://www.repairmanuals.com to find the specific *Chilton's Repair Manuals* for your vehicle. If your vehicle is not in *Chilton's Repair Manuals*, try the *Haynes' Repair Manuals*. You can go to the same website for both manuals. The manuals explain in detail domestic and foreign vehicle repairs. If you cannot find your vehicle in these repair manuals, you can go to the automotive discount stores. They will be glad to help you.

Let's get serious here! Checking brakes is a very simple job to do. You have to take the time to remove the tires. Take the tires off your vehicle one at a time. After you remove one of the tires, check the condition of the *disc rotor brakes* (on the front) and the *drum brakes* (on the back) wheels of each one of the front brakes and the back brakes. The *drum brakes* are on the inside of the back *brake drums*. That is how the back *drum brake* got its name. What should you be looking for? While wearing a pair of mechanic's gloves, take your finger and lightly touch or look at the shiny part of the disc rotor brake or inside the drum brake (where the brake drum is located). It should feel or look smooth. Do you feel or see deep ridges with very sharp edges? If the answer is "yes," your vehicle will have worn or damaged *brake pads* and *brake shoes*. The *brake pads* and *brake shoes* need to be replaced.

What are *brake pads* and *brake shoes*? *Brake Pads are a part of the disc rotor brake.* The brake pads are glued onto a steel plate. The brake pads make contact with the front disc rotor brakes to help the front tires to slow down or stop when you press your foot on the brake pedal. The *brake pads* wear with use.

Brake Shoes are inside a drum brake. They are made of steel which is in *the shape of a shoe.* The brake shoes have brake material glued to them. The brake shoes expand outward and come in contact with the *brake drum* to help the back tires to slow down or stop when you press your foot on the brake pedal. The *brake shoes* wear with use. The back drum brakes are used when you engage the parking brake on steep hills or inclines. The *parking brake* and the back drum brakes stop your vehicle from rolling forward or backward down the hill or incline.

Today, most of the newer vehicles are being made with disc rotor brakes and drum brakes that are thin. The mechanic may not be able to "machine" the disc rotor brakes or the drum brakes to get a smooth surface on them again, so they can be used again on your vehicle. ("The machining of the disc rotor brakes and the drum brakes is called turning the rotors or drums.") Usually turning the disc rotor brakes and the drum brakes are cheaper than buying new disc rotor brakes and drum brakes. It would be easier and less of a headache if you allow a mechanic to "machine" the disc rotor brakes and the drum brakes for you.

A good mechanic will tell you if he disc rotor brakes and the drum brakes need to be replaced, and if he can "machine" the old ones to a safe tolerance to save you some money. The thickness of the disc rotor brakes and the drum brakes will determine whether they can be reused after "machining." Your vehicle needs its disc rotor brakes and drum bakes to be a specific thickness to allow you to slow down or stop your vehicle safely. It is a very good idea to change the *disc rotor brakes* and the *brake pads* (on the wheels of the front tires of your vehicle's braking system) because they may be too thin to work properly. I always replace the *disc rotor brakes, brake pads,* and *brake calipers* when I get the worn brakes changed on my Ford Windstar.

A brake caliper is the piece of metal that fits over the back of the wheel of the disc rotor brakes on the front braking system of a vehicle. The photograph at the beginning of this chapter is a disc rotor brake with a brake caliper on the back of it. The brake pads are on the inside of the *brake calipers*. The *brake calipers* squeeze the disc rotor brakes when you press your foot on the brake pedal to slow down or stop the front tires of your vehicle.

The *drum brakes* are on the back wheels of a vehicle's braking system. The drum brakes need to be changed also when they are worn. Inside a drum brake are *brake shoes* and other parts that help the back tires of your vehicle to slow down or stop when your press your foot on the brake pedal. The photograph below is a drum brake. The *parking brake* in a vehicle only works with the drum brakes and only stops the back tires. Almost all vehicle manufactures (including Ford) go the cheapest route possible to make parts for different types of vehicles including recreational vehicles. The parts wear out with use over time.

Anytime you apply your brakes while you are driving and you hear a sound that sounds like grinding metal, it might be the *wear indicators* on the brake pads and brake shoes on the wheels of braking system of your vehicle. Some brake manufactures do not put *wear indicators* on their brakes. If your vehicle has *wear indicators* on the brakes, you *will hear a grinding metal sound when the brakes are worn and unsafe to use on your vehicle*. The *wear indicators* are located on the brake pads (on the front) and on the brake shoes (on the back) of your vehicle's braking system. The *wear indicators* are there to alert you that your brakes are worn and need to be replaced.

While your vehicle's tires are removed, it would be a good idea to check for any leaking brake fluid from the brake fluid lines around your vehicle's brakes. Leaking brake fluid means your vehicle will lose its braking power very quickly. If you do not fix the brake leak, you could crash into something. Checking for leaks will prevent any braking problems from happening. That means you will have brakes when you need them!

Here is a problem you do not want. What are *spongy brakes*? *When the brake fluid level gets to low, air will get into the braking system and will cause spongy brakes*. When your vehicle has *spongy brakes*, you have to keep pumping the brake pedal to get your vehicle's brakes to work. Your vehicle will not be able to slow down or stop quickly. Worst of all—your vehicle will not be able to stop. The brakes have failed on your vehicle.

Sometimes you might need to add brake fluid to your vehicle's braking system. You need to find the *master brake reservoir*.

Usually, you will find it under the hood of your vehicle in the engine well. Look for a panel behind the engine, this is the firewall. The brake fluid reservoir should be mounted to the firewall. There are two marks on the left side of the brake fluid reservoir, with the abbreviations for the words "minimum (min.)" and "maximum (max.)" brake fluid levels.

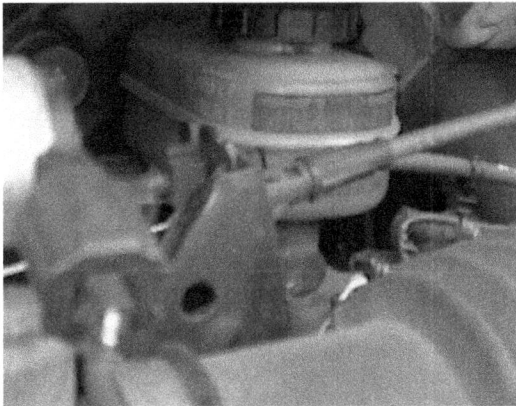

In the photograph, you can see the brake fluid reservoir in my mini-van mounted to the back firewall. You can see the "minimum and maximum level" marks on the left side of the brake fluid reservoir below the filler cap (on the top of the brake fluid reservoir). You are learning the very basics of a braking system on a vehicle.

Use brake fluid in the brake fluid reservoir. Do not use water. The brake fluid will not freeze during the winter. You do not want your vehicle's brakes to fail in wet, rainy weather, or icy, snowy weather conditions. You also need your vehicle's brakes on slick roads or highways. Oil spills, chemical spills, and other hazards can cause slick dangerous road and highway conditions. Do not over fill the brake fluid reservoir with brake fluid. It will cause seals and lines to blow and start leaking. The brake fluid reservoir has a vented cap on it. This will help get rid of some of the pressure in the brake fluid lines, but the vented cap will not stop brake fluid leakage from other parts of the braking system. You can buy brake fluid from vehicle dealerships and automotive discount stores.

Many brake fluid bottles have a center pour spout. To pour brake fluid from a center pour spout, you have to angle the bottle at an upward slant and pour very slowly to allow air to get into the bottle to help the brake fluid to flow more evenly. Use a funnel to help you pour the brake fluid into the brake fluid reservoir without making a mess.

Take the cap off the brake fluid reservoir and pour in enough brake fluid to the "maximum (max.) level" mark, so the brakes on your vehicle will be able to work. Drive your vehicle a short distance to test your vehicle's braking and stopping power. Check your vehicle's brakes once-a-month.

Buying a vehicle is an expensive investment. Doing some of the preventive maintenance on your vehicle is a great way to learn to take care of your vehicle. Aren't maintenance tips fun and money saving?

Chapter 7

Power Steering: Going Right or Left

Power Steering is a very nice thing to have in any vehicle. Can you imagine using a pair of ropes to steer a car or truck in traffic? *Power Steering or power-assisted steering as it is called in the automotive manufacturing industry* is very easy to understand. Today, *power-assisted steering* is easier to operate and more reliable than the manual steering vehicles of the past. I remember driving older cars with manual steering. Those older cars were not fun to drive. The manual steering cars were hard to steer at low speeds. It was hard to turn the steering wheel when I was backing up, parking, turning the steering wheel while the car was standing still, or turning corners. I am going to explain to you and show you how uncomplicated it is to maintain the *power steering* on your vehicle. I am going to teach you the basics of a power steering system (or power-assisted steering system). Let's start with the parts that you can find in your car, truck, or mini-van. The basic *power steering system consists of a pump and a serpentine belt that is driven by your vehicle's engine*. The power steering pump is connected to an engine's serpentine belt or an engine's V-belt (or engine drive-belt). Check your owner's manual for the correct name of the engine belt on your vehicle.

The power steering unit has two lines. One line goes from the power steering pump to the power steering unit, and the second line((power steering fluid return line) goes from the power steering unit back to the power steering pump. *The power steering unit in most vehicles today, consists of a powered rack and pinion unit.* The *powered rack and pinion unit* is connected to the front wheels of your vehicle by *tie-rods*. This is what moves your vehicle's front wheels in the direction of right or left. *The back wheels are fixed. This means the back wheels follow the front wheels.* The back wheels have no steering capability. When the manufactures talk about the wheels on a vehicle, they are talking about the tires and tire rims as a single unit.

Open the hood of your vehicle to see how your vehicle's *power steering* operates. While the engine is running at an idle and not running at full throttle, the *power steering pump* is spinning being connected to the *serpentine belt*. The *serpentine belt* is driven by the engine. As the *power steering pump* is spinning, it pumps power steering fluid under pressure through the two lines to the *powered rack and pinion unit* helping you to steer your vehicle's tires with less effort on your part. By turning the steering wheel with the help of *the powered-assisted steering system*, the wheels on your vehicle have no problem turning in the direction of left or right. Power Steering Fluid returns from the *powered rack and pinion steering unit back to the power steering pump and the power steering fluid reservoir*. Some vehicles have the power steering pump and the reservoir as one unit. On my Ford Windstar, the power steering pump and the power steering fluid reservoir are two separate parts of the *power-assisted steering system*.

There are two types of reservoirs for the power steering fluid. The power steering fluid reservoir in my Ford Windstar has only line markings on the side of the reservoir. The reservoir has three line marks on it. The top line is the *maximum* fluid level of power steering fluid the reservoir will hold. The middle line from the top line is considered the *acceptable* fluid level for the power steering fluid. The bottom line is the *minimum* fluid level of power steering fluid remaining in the reservoir. For the *power-assisted steering* to work at its best, the reservoir must be filled with power steering fluid to the top line (or *maximum*) mark The *power-assisted steering* will still work at the *acceptable* or middle line mark, but the power steering fluid should always be at the top line (or *maximum*) mark to work at its optimal level.

The second type of power steering fluid reservoir has a *dipstick* permanently attached to the inside of the cap. The *dipstick* has a *maximum level* mark at the top and a *minimum level* mark at the bottom of the *dipstick*. The diamond or *hatch marks* in the middle of the *dipstick* is considered the *acceptable level* mark for power steering fluid in the power steering fluid reservoir.

The basic care of the *power-assisted steering system* is uncomplicated. The **only time** you can check the power steering fluid in your vehicle is when the engine is running at an idle and parked. If you check the power steering fluid while the engine turned "off," you will not get an accurate reading on the amount of power steering fluid remaining in the reservoir. You need to check the power steering fluid level in your vehicle at least once-a-week. Check the two lines for leaks too.

This is the easiest way to check the power steering fluid lines for leaks. Start your vehicle's engine while it is in park running at an idle. Open the hood of your vehicle. Turn the steering wheel all the way to the right until you cannot turn the steering wheel any more. This is called a "wheel-stop." Now, turn the steering wheel all the way to the left until the steering wheel stops turning. Next, turn the steering wheel so that the front wheels of the vehicle are straight (or facing forward). Get out of your vehicle, and look under the hood at the lines attached to the *power steering pump* and the *power steering reservoir*. Are there any signs of leakage? You will see a fluid that is red in color or a fluid that is clear (without color) on the outside of the lines if there are leaks. The lines will look wet, and there will be a wet area around the *power steering reservoir*.

If you hear noises while you are driving, and it becomes hard for you to steer when your vehicle makes turn; these are good indictors that your *power steering system* is low in power steering fluid. **If there is very little power steering fluid in the power steering system, the power steering pump will fail (or burn up).** Power Steering Fluid makes the power steering system work, and it is a lubricant for your vehicle's power steering system.

Do not over fill the power steering reservoir. (I am pointing to the reservoir in the back.) Excess fluid could blow the power steering lines and cause the loss of the power steering pump because of lack of lubrication. I advise very strongly that you refer to your owner's manual if you have any questions about the *power steering system* in your vehicle. Use your owner's manual for finding the correct type of power steering fluid needed for your vehicle to fill the power steering reservoir. For example, my Ford Windstar uses a Mercon ATF for its *power steering system*. ATF is an acronym for **A**utomatic **T**ransmission **F**luid. Remember, if in doubt; always refer to your owner's manual for the type of power steering fluid needed for your vehicle.

There are two types of power steering fluid. The first type of fluid is red in color. The automotive manufacturers use this type of fluid. It is transmission fluid. The second type of power steering fluid is clear (without color). It is an after-market power steering fluid. This is a generic type of power steering fluid that can be used in most vehicles. You can find both, the red in color and clear (without color) power steering fluids, at automotive discount stores, department stores, vehicle dealerships, and gas stations.

Believe it or not, my power steering system never gave me any problems to tell you about since I regularly maintained it. I have never had any steering problems since I bought my 2000 Ford Windstar. I hope you now have a better understanding of your vehicle's *power steering system* and how to care for it.

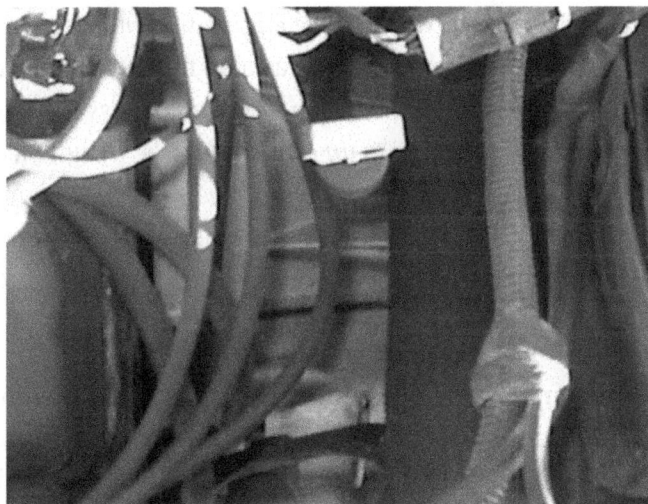

Chapter 8

Transmissions and Transaxles:

Go, Go, Go

Ever since a car, truck, or mini-van was built it has had a *rear wheel drive transmission*. Transmissions have changed greatly in the last thirty years. I would like to give you a very brief history about transaxles to help in your understanding of this chapter. In the year 1977, a car company named Volkswagen in Germany, introduced a car called the *Rabbit*. It featured a rather unusual *drive system*. It was a front wheel drive car. The engine was mounted in a *traverse position. (The engine was mounted across the engine well.)* The *transmission* was also mounted in a *traverse position. (The transmission was also mounted across the engine well, and it had a final drive train attached to the front wheels.)* In the year 1977, this kind of *driveline or drive train* was completely different for a car. Three years later the American automotive industry started building front wheel drive cars like the Volkswagen *Rabbit*.

The two main reasons the automotive industry went with the *front wheel drive vehicles* in the 1980's are as follows: First, the manufactures gained more interior room in their cars and mini-vans by moving the drive shaft to the front wheels. Second, cars and mini-vans now have most of the weight of the vehicles on the front wheels of the cars and mini-vans. The *front wheel drive system* causes cars and mini-vans to have more traction and handle very well in bad weather conditions. Another benefit of a *front wheel drive system* is the vehicles have more control and better steering of the front wheels in bad weather. The majority of the weight of the cars and mini-vans is in the front of these vehicles because of an engine and a transaxle used by the automotive industry.

Performance Cars, full-size vans, smaller trucks, and full-size trucks are the only exception. Some of the automotive manufacturers still build these vehicles with *rear wheel drive and an older traditional transmission.*

To determine whether you have a *front wheel drive or rear wheel drive vehicle*, it will depend on the make, model, and year of the vehicle. What is a *rear wheel drive vehicle*? This means that *the transmission moves the rear wheels forward at different speeds with gears and moves the rear wheels in reverse with a gear*. The *rear wheel drive* moves the vehicle forward and in reverse. The front wheels of a *rear wheel drive vehicle* just steer the vehicle.

Today, most new vehicles use a *transaxle* type of *transmission*. *A transaxle is a combination of a transmission and a final drive train, or it is also called rear end drive. A final drive train (or a rear end drive) attaches the transaxle to the front wheels.* One, small, compact housing holds the transmission in a *rear wheel drive vehicle* or the transaxle in a *front wheel drive vehicle*. The housing is space-saving for *rear wheel drive* and *front wheel drive vehicles*. My Ford Windstar is a *front wheel drive vehicle*. The *front wheel drive vehicles* handle better than the *rear wheel drive vehicles* in snow, ice, and rain. It surprises me how well my Windstar handles in bad weather. It does not matter whether your vehicle has a *rear wheel drive system* or a *front wheel drive system*. The advice I give you for maintaining a car, truck, or van will still help you in maintain your vehicle.

There are two ways to operate *rear wheel drive transmissions* and *front wheel drive transmissions*. Transmissions can operate manually or automatically. The automotive industry uses certain criteria before deciding whether a vehicle will have an automatic or a manual transmission installed. The automotive industry looks at the type of vehicle being built, the make, model, year, what the vehicle will be used for, production costs, how to reduce production costs, what sells and appeals to customers, and many other criteria are used in deciding whether a vehicle will have a manual or automatic transmission.

The first type of operating system for transmissions is a manual transmission, and the second type of operating system for transmissions is an automatic transmission. A manual transmission is simple. You do all of the shifting for the vehicle to go forward or in reverse. The manual transmission in a vehicle needs to be maintained for it to operate properly. You need to check the oil level or automatic transmission fluid level in a manual transmission.

There are two types of manual transmissions. The traditional manual transmissions (with *rear wheel drive*) use a heavy weight gear oil most of the time. The gear oil is a 90-*weight* gear oil (Gear Oil *90 W*). *The number 90 represents the thickness and the viscosity of the gear oil.* In the automotive industry, the word "viscosity" means "resistant to flow." The gear oil needs to stay on the gears of the manual transmission to keep the transmission working.

In the early 1990's, the traditional *rear wheel drive vehicles* with manual transmissions were built more efficiently by the automotive industry. The government of the United States imposed fuel mileage regulations that required them to change. The automotive industry had to replace the heavyweight gear oil with automatic transmission fluids. Today, the automotive industry is building smaller engines and transmissions in performance cars, full-size vans, small trucks, and full-size trucks. The automatic transmission fluids are thinner and have a lower "viscosity" *number (of 20 to 40)*. The automatic transmission fluids flow very easily and allow more horsepower to go to the *rear wheel drive*, which allows the gears to work more effectively. This increases the fuel mileage of your vehicle.

The second type of manual transmission is a manual transaxle transmission with *front wheel drive*. This type of transmission has been in manual vehicles since the 1980's. The automotive industry is building smaller engines and transmissions in cars and mini-vans. The transaxle transmission is a new design for manual transmissions that use thinner automatic transmission fluids with a lower "viscosity" *number of 20 to 40* to make an engine work less. The thinner fluids give more horsepower to the transmission for the working of the gears. This gives your vehicle better fuel mileage. You need to check the owner's manual of your vehicle to find the correct type of automatic transmission fluid or the correct weight of the oil needed for your vehicle.

The easiest way to check a manual transmission in a vehicle is by noticing how easy it is to shift from one gear to the next gear. If a vehicle's manual transmission has difficulty shifting, the vehicle might have damaged gears or damaged *synchronizers* in its transmission. The most common reason for a manual transmission to have problems shifting from one gear to another gear is low transmission fluid or oil. You forgot to check the level of the transmission fluid or oil in your vehicle. Some owners' manuals might show them how to check the fluid level on the manual transmissions of their vehicles.

To check the fluid level of a manual transmission, look for two plugs on the side of the transmission underneath your vehicle. One plug will be higher than the other. The *upper plug* is called an *inspection plug*. You may be able to get to the *inspection plug* on some vehicles by opening the hood of your vehicle. The second or *lower plug* is called a *drain plug*. The *drain plug* will always be underneath a vehicle. The *inspection plug is where the transmission oils and automatic transmission fluids are added.* Remember to check your owner's manual for the correct fluid or oil for your vehicle. Your owner's manual may help you to locate the *inspection plug* and the *drain plug*. To remove the threaded *inspection plug*, turn the plug counter-clockwise (turn the plug to the left) with an open-end wrench, and it should come off easily. If the threaded *inspection plug* does not come off counter-clockwise, then, turn the plug clockwise (turn the plug to the right) with an open-end wrench. Check the level of the transmission fluid by putting your finger down inside the hole where the plug was. The transmission fluid should be at the bottom of the hole when you put your finger down inside the hole. (When you stick your finger down inside the hole, you should feel the transmission fluid.) If the hole is filled with the correct transmission fluid or oil and is at the correct level, clean the plug; then, put the plug back into the inspection plug hole in the transmission, and tighten down the plug in the opposite direction with an open-end wrench. There is no *dipstick* for the *inspection plug* on manual transmissions. If you do not want to get your hands dirty, you can pay a mechanic to stick his finger in the inspection plug hole to check the transmission fluid level of the manual transmission in your vehicle. This is the way it is done with older vehicles and some of the newer vehicles today.

Using the wrong type of oil or automatic transmission fluid in your vehicle's manual transmission is the same as running with low fluid or no fluid at all in your vehicle's transmission.

This is how you put manual transmission oils or automatic transmission fluids in a manual transmission vehicle. You need to buy a funnel with a tube on the end of it. You can find these funnels at automotive discount stores or Wal-Mart stores. Pour the fluid down inside the inspection plug hole until the fluid reaches the bottom of the plug hole. Put the cleaned plug back into the hole, and tighten it down with your open-end wrench.

The *lower plug* on a manual transmission is called a *drain plug*. This plug is *used for draining transmission oil or automatic transmission fluid out of a transmission before it has to be repaired or replaced*. Before you drain the oil or automatic transmission fluid out of your vehicle's transmission, drive your vehicle *20* to *30 minutes*. The warm fluids will drain easier and faster and parking on level ground will help drain as much fluid as possible from your vehicle's transmission.

Use a *drip pan* to collect all of the transmission fluid. You can buy a *drip pan* at automotive discount stores or Wal-Mart stores. **Do not dispose of the transmission oils or fluids in sewers, storm drains, any body of water, or landfills that will cause a public health hazard. A mechanic's shop or a dealership can help you dispose of the transmission fluids.**

Manual Clutches are another thing that can be easily checked with your vehicle's engine running at an idle. Do not run the engine of your vehicle at a full throttle. When you release the clutch, no matter what gear your vehicle is in; reverse or another gear you will be hurled, to move with force, forward very quickly at a high rate of speed into a wall or other immoveable objects around you. There could be a lot of property damage and loss of life, including yours.

This is the way to check a manual clutch. Engage the clutch of your vehicle. Clutch and shift into first gear, then, slowly release the clutch and watch how easy your vehicle moves forward while the engine is running at an idle. If your vehicle moves slowly without jumping or shaking, and it does not have any other problems when moving in any direction; your vehicle's clutch is operating correctly. The jumping or shaking action of your vehicle means that the clutch is either worn, or it has oil on it, or it needs adjustment to work properly. Look for oil leaks since the engine's rear oil seal is near the clutch.

What is a *throw out bearing? A throw out bearing is a device that puts pressure on a pressure plate to release the clutch to allow shifting into any gear.* To keep a manual clutch working, you must check the *throw out bearing*.

This is the best method to check a *throw out bearing*. Listen for a whirling (to move or spin rapidly) high pitched sound. Do you hear a high pitched sound before you depress the clutch to shift and the sound disappears and then, the bearing sound comes back after using the clutch (after shifting into any gear)? If the answer is "yes," the *throw out bearing* is worn and needs to be replaced.

If you do not have time, this is a quick method to check a *throw out bearing*. Slowly depress the clutch a quarter of the way to the floor, then, slowly release the clutch and listen for any whirling sound. Do not abuse the clutch or transmission in your vehicle. Your vehicle will absolutely stop working. You will be stuck somewhere, a town, a city, a deserted highway, or some place you do not want to be. The upkeep and care of a manual transmission vehicle takes more effort on the owner's part than an owner of a vehicle with an automatic transmission. I do not own a vehicle with a manual transmission. You will probably wish you had an automatic transmission or automatic transaxle in your vehicle.

For those of us who drive vehicles with traditional automatic transmissions and automatic transaxles, here is some information you need to know. Automatic Transaxles are more common in *front wheel drive* cars and mini-vans. Full-size trucks, small trucks, and full-size vans with *rear wheel drive* have a traditional automatic transmission.

Since we are on the subject of automatic transmissions, putting the correct type of fluid in your vehicle is of utmost importance. For example, the transaxle in my2000 Ford Windstar uses a Mercon V (Mercon 5) type of fluid .There is another type of transmission fluid called Mercon III (Mercon 3) that is used for older Ford Windstar mini-vans. When you mix Mercon V (Mercon 5) and Mercon III (Mercon3) together, **you will have a very thick transmission fluid that will not work in an automatic transmission or automatic transaxle. Mixing these two transmission fluids causes poor shifting or no shifting at all.** I hope I am making this very clear to you. If you ask, what does this all mean? The automotive industry after years of building automatic transmissions has developed automatic transmission fluids that make the transmissions work with the least amount of effort by the engine. Automatic Transmissions are huge horsepower hogs on an engine; it tales quite a bit of engine power to make a vehicle go down the road. It makes sense to develop an automatic transmission fluid that reduces the horsepower requirements to make the transmission work more efficiently. Refer to your owner's manual if you are not sure what type of transmission fluid is used in your vehicle.

Some automatic transmission vehicles do not have *dipsticks*. The *dipstick* has been replaced with a *dummy plug*. The *dummy plug* is on the right or left side of the transmission near the *inspection pan*. You will have to get underneath your vehicle, or go to a dealership, or a mechanic's shop to check the level of the automatic transmission fluid in your vehicle's transmission. Ford decided to lower its costs by using a *dummy plug* instead of a *dipstick*. I do not like this idea. I can never tell whether my vehicle's transmission fluid level is correct. I do not know if other automotive manufacturers use a *dummy plug*. Check your owner's manual for more information.

You can find the *dipstick* of a traditional automatic transmission on the driver's side of the vehicle under the hood of the vehicle. This allows you easy access to check the transmission fluid level. A *dipstick* is not always found under the hood of a traditional automatic transmission vehicle. Check your owner's manual for the specific location of the *dipstick*. The *dipstick* could be underneath the vehicle.

Look for *dipstick* under the hood of your vehicle. The *dipstick* on an automatic transaxle transmission is usually on the top side of the transaxle housing. This allows you easy access to check the transmission fluid level. Always refer to your owner's manual for the specific location of the automatic transaxle transmission dipstick.

Some automatic transmission vehicles do have *dipsticks* check the transmission fluid level of the vehicles. You need to know how to read the *dipstick* of an automatic transmission. You need to know how to check the fluid level of an automatic transmission. This is very to do. Let me walk you through it. The first thing to keep in mind is **never** check a transmission's fluid level when the engine is **cold**. Make sure your vehicle's engine is warmed up and running at an idle. The primary reason for this is, transmission fluids expand when they are warmed up. For example, after you have driven your vehicle *20 miles or more*, it should be warmed up enough to check the automatic transmission's fluid level. Do not be like the guys in the gas stations, who check the automatic transmission fluid level of their vehicles while pumping gas with the engine turned "off." They will not get an accurate reading from the *dipstick* on the amount of automatic transmission fluid in the transmissions of their vehicles. They will be driving their vehicles on low transmission fluids. Automatic Transmission Fluids expand when they are warmed up. This gives you an accurate reading of the amount of fluid in your vehicle's transmission while it is running at an idle. If you add any fluid to an automatic transmission when it is cold, you run the risk of over filling the automatic transmission. The fluid will have no place to go when it heats up and expands. As the fluid heats up and expands, it fills up empty spaces in a transmission. The excessive fluid will cause gaskets and seals on the transmission to leak or "blow out." (A seal or gasket on the transmission gets pushed out of place.) You will have to replace the seals and gaskets on the transmission. The transmission in your vehicle could die. You will be saying, "My vehicle's poor transmission rest in peace!"

It is easy to read the *dipstick* on an automatic transmission. The *dipstick* on an automatic transmission has similarities and differences to an engine oil dipstick. The two most important similarities are as follows: First, the *dipstick* is designed just like the engine oil dipstick in your vehicle. The *dipstick* has a *minimum level* and a *maximum level* marked on it for transmission fluids. You can see the two holes (that look like black dots in the photo-graph below) that mark the *maximum level* and *minimum level* of transmission fluid to be added to your vehicle's transmission. There is a *cross-hatching* section also on the *dipstick* of the transmission.

The *cross-hatching* section on the *dipstick* shows the *acceptable* level for transmission fluid in your vehicle. The second similarity is that you have to use a specific type of automatic transmission fluid for the transmission in your vehicle to operate properly. This also is similar to the different weights of oils used in the engine of your vehicle.

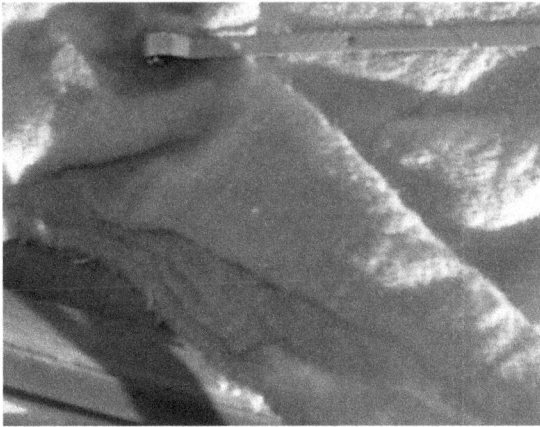

There are differences between an engine oil dipstick and an automatic transmission dipstick. The two main differences are as follows: First, engine oil levels are checked when the engine is turned "off" for an accurate reading. Transmission Fluid is checked when an engine is running at an idle for an accurate reading. The second difference is that the *cross-hatching* section on an *engine oil dipstick represents a vehicle being one quart low on engine oil.* The *cross-hatching* section of an *automatic transmission dipstick represents a vehicle that is one quart or greater than one quart low on transmission fluid.*

Here is how you add transmission fluid to the transmission. If you need to add transmission fluid to your vehicle, add it in half-pint increments. The reason the transmission fluid is added to a traditional automatic transmission or automatic transaxle in half-pint increments is to prevent overfilling of the transmission. Pour the transmission fluid into the *dipstick tube. (The place where you pulled out the transmission dipstick is called a tube.)* Use an ordinary funnel to help you pour the fluid accurately and carefully without a mess. Please remember to check your owner's manual for the specific fluid for the transmission in your vehicle.

You need to wait for the fluid to settle before checking the fluid level again. There are two ways to do this. One way to do this is as follows: You can wait *20* to *30 minutes* before you check the transmission fluid level in your vehicle again. Another way to check the fluid level is as follows: It would be a good idea to drive around for a few miles, and then, check the fluid level again by pulling out the *transmission dipstick.* Keep doing this until the fluid level is *above* the *cross-hatching* section of the *dipstick,* which *indicates the acceptable operating range of the transmission.*

Here is basic information on how a traditional automatic transmission or automatic transaxle operates. The design of the transmission could have a *torque converter* or it is called a *fluid pump* and *a gear pump,* or the design of the transmission could have only a *torque converter (fluid pump).* Mechanics call the *device between the engine and the transmission a torque converter or fluid pump.* This device uses fluid connection for the engine to drive the transmission without any direct connection to the transmission. The easiest way to understand how the *torque converter* or *fluid pump* operates is—get two electric fans. Place the two fans facing each other, plug one fan into a wall outlet and it "on." Watch how the air flow causes the blades to move on the second fan. The blades spin without the second fan being plugged into a wall outlet. This is how a *torque converter* or *fluid pump* works with an automatic transmission fluid to make the transmission operate properly. This is why you can sit at a stop light waiting for the light to change or waiting for the traffic to pass while your vehicle is in drive without using a clutch to disengage a gear. The transmission fluid is under pressure and passes through the internal transmission *gear pump* and *torque converter* or *fluid pump*. This makes shifting possible inside of a transmission.

What controls the transmission shifting in your vehicle? The valves and a device called a "governor" inside of the transmission. A "governor" is "a device that tells the transmission when to shift." The "governor" does this by the vehicle's speed and the revolutions per minute (or the r.p.m.) of the vehicle's engine.

How does the automatic transmission in your vehicle shift from one gear to another gear? The "governor" "controls the opening and closing of the shifting bands inside the transmission." Planetary Gear Sets are controlled by the shifting bands. As the shifting bands open and close around the planetary gear sets, this allows the transmission to shift into different gears. Now, you can shift your vehicle into drive, reverse, first gear, second gear, or third gear (called overdrive). See how important it is to keep your vehicle's automatic transmission fluid level in the acceptable operating range.

Sometimes transmissions have problems like *Hard-shifting*. It happens when you are shifting to make your vehicle go forward in the drive position on the steering column. Your *vehicle will suddenly bolt forward very fast.* It is like getting a kick in the pants. Have you ever wondered why automatic transmissions do not want to shift into any gear or are slow to shift to get your vehicle moving? The most common reasons are a low transmission fluid level or dirty transmission fluid in the transmission. Now, you can see what an important and complex piece of machinery an automatic transmission is sitting inside of a vehicle.

Ask yourself these questions. Do you hear funny sounds coming from your vehicle? Do the noises sound like a whining or grinding noise from your vehicle's transmission when you are driving? If the answer is "yes," this is a good indication that your vehicle has serious problems inside of its transmission. The main cause of most serious transmission problems is low transmission fluid. You need to check the fluid level at least once-a-week, and keep the transmission fluid level in the *maximum* operating range (above the *cross-hatching* section) on the *dipstick* to prevent transmission problems. Vehicles with *dummy plugs* should have the transmission fluid level in the vehicles checked every 3,000 to 5,000 miles when you do an engine oil change in your vehicle.

Keep this in mind! Automatic Transmissions and Automatic Transaxles need to have the fluid changed; just like the engine oil in your vehicle. Transmissions that are built today require a fluid change at 30,000, 60,000, 90,000, and 120,000 miles on the odometer or every 30,000 miles.

There is a filter in a traditional automatic transmission and an automatic transaxle. Do not forget to change the transmission filter when you change the fluid in your vehicle's transmission. The purpose of the transmission filter is to catch the small pieces of metal, metal shavings, and shifting band material that wear out and break during the operation of a transmission. The filter helps to keep the transmission fluid clean, and it prevents transmission problems caused by dirty transmission fluid.

You will have to get under your vehicle to do this job yourself, or you can hire a mechanic to do it for you. The filter for an automatic transmission or an automatic transaxle is located on the lowest part of the transmission in an area called the transmission *inspection pan*. The transmission filter is inside the pan. You will have to drain the transmission fluid out of the transmission before you can get access to the filter. A transmission can be drained by using the *transmission drain plug*. The plug is located on the transmission *inspection pan*. Next, remove the pan to get full access to the filter. Pull off the old filter, and replace it with a new filter. You just have to push the new filter in place. Do not forget to snuggly tighten down the bolts on the transmission *inspection pan*. You do not want to see any leakage from the pan. **Do not dispose of the transmission fluids in sewers, storm drains, any body of water, or landfills that will cause a public health hazard. Dispose of the transmission fluids at a mechanic's shop or a dealership.**

By doing automatic transmission fluid changes and changing the transmission filter, you are keeping your transmission running longer and your wallet happy. See how simple it is to car for an automatic transmission. You did not need a hammer and a chisel to do this job.

For those who are interested in some facts and to give you more to think about, the original transaxle on my 2000 Ford Windstar stopped working at 441,919 miles on the odometer. That is almost half-a-million miles. Who said vehicle maintenance does not matter? My mini-van's transaxle lasted this long simply by watching its fluid level and doing fluid and filter changes at or about 100,000 miles and when I could afford doing the changes. I also changed the fluid and filter at different times before the 100,000 miles on the odometer. For example, when the transaxle was not shifting correctly in each gear, I knew the transaxle fluid and filter needed to be changed. I drove my Ford Windstar sanely! You should have seen the mechanics' faces at the Don Meyer Ford and Lincoln dealership, in Greensburg, Indiana when they saw the mileage on my mini-van's odometer. I told them it was the original transaxle that came with the mini-van. The original transaxle on my mini-van stopped working at 441,919 miles, but it was still able to be rebuilt. The only thing that stopped working was the *torque converter* which caused the transaxle to fail. Keep your vehicle's transmission happy by taking care of it. You will have fewer headaches.

Chapter 9

Battery: Charge it Up

In this chapter I am going teach you about a vehicle's battery. There are basic things you need to understand about the battery in your vehicle. The vehicle batteries of today have become "sealed." This means you "do not have to add distilled water or battery acid to a battery. " Today, batteries require very little maintenance. Before the mid-1970's, the automotive industry did not have *sealed batteries*. It put batteries in vehicles that came with a removable vent cap. You had to add the distilled water and the battery acid to maintain the battery. You had to check the distilled water level and battery acid level in the battery with a tool called a *battery hydrometer*. The *hydrometer checked the gravity of the distilled water*. This means *the amount of battery acid mixed with the distilled water*. You had to do this for all six cells on the battery. Hurry, for modern technology.

Look for the battery of your vehicle under the hood of your vehicle. The battery should be on the right or left side of the vehicle. Your owner's manual will have pictures to show you where your vehicle's battery can be located.

Most batteries today have a little *sight glass* about the size of a dime on top of a battery to view the condition of the battery. The *sight glass* on the battery will show any one of these three conditions of the battery. There are three colors on a battery to indicate the condition of the battery. This is important to know. A *good battery* is the color *green* in the *sight glass*. A *battery* that is *having problems* is the color *yellow,* and a *dead battery* is the color *red.* For example, the battery in my Ford Windstar is a "sealed" type of battery with vent caps. The *sight glass* would be under the carry strap of the battery. You have to move the carry strap to see the *sight glass*. Not all vehicles have a carry strap on the battery. My 2000 Ford Windstar has a carry strap on the battery, but it does not have a *sight glass*. The reason some batteries do not have a *sight glass* is, batteries are rated in months. The battery industry rating for batteries is 36 months to 100 months. For example, my Windstar has an 84-month battery in it. The battery lasted the full 84 months before it died. The photograph on the next page shows you a *sight glass*, but it is not from my mini-van.

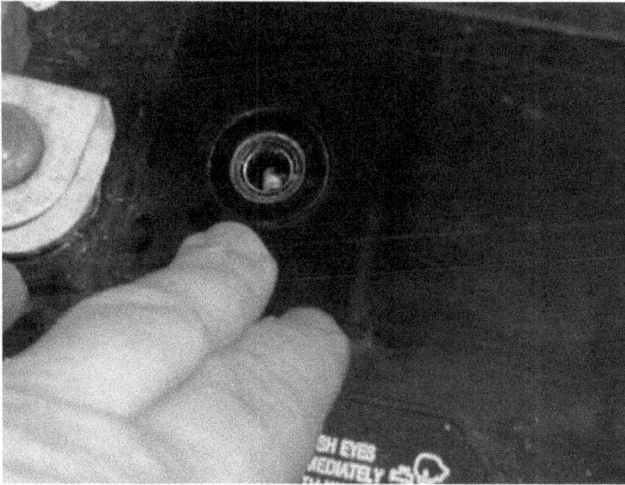

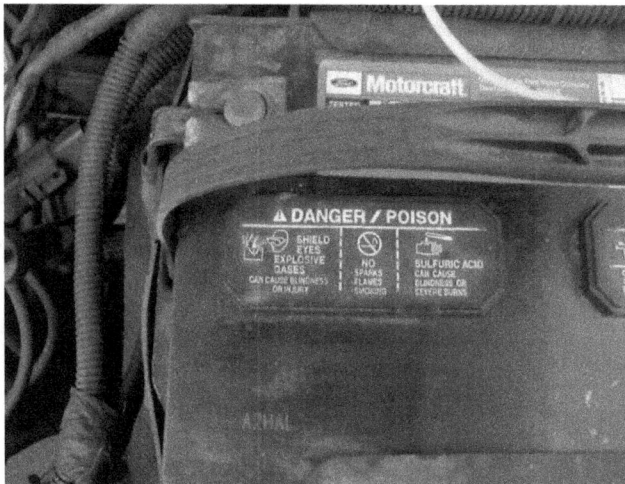

79

The *vent caps* cover the six cells of the battery. (There are three cells under each cap. The photographs in this chapter show two raised areas, one on either side of the battery under a carry strap. The *vent caps* have *important safety instructions written on them.*) **Do not remove** the *vent caps* from the battery for any reason. Why do batteries have *vent caps*? As a battery is charging and discharging, it produces hydrogen gas inside the battery. *The hydrogen gas needs a way to escape (or vent out) of the battery.* This prevents the excess pressure from the hydrogen gas from damaging the battery. Follow the safety instructions on the battery. **Do not smoke a cigarette or other products around a battery.** No smoking around a battery prevents the chance of an explosion. The hydrogen gas from the battery is **very flammable** and **will ignite.**

You need to take care of your vehicle's battery to help it last longer. What steps should be taken to maintain a battery? Batteries

80

need to be cleaned about every six month. You can clean the outside of a battery with a clean and dry rag to get all of the dust and loose dirt off the battery. The battery posts also need to be cleaned with a tool called a *battery posts cleaner*. When you see rust and a white powder (lead oxide) around the battery posts, it is time to clean the battery posts terminals. You need to wear gloves to prevent the lead oxide and sulfuric acid from getting onto your hands from the battery. You can buy a pair of leather work gloves or a pair of general purpose mechanic's gloves at any automotive discount store. (See the photograph below. The leather work gloves are on the left. The general purpose mechanic's gloves are on the right.)

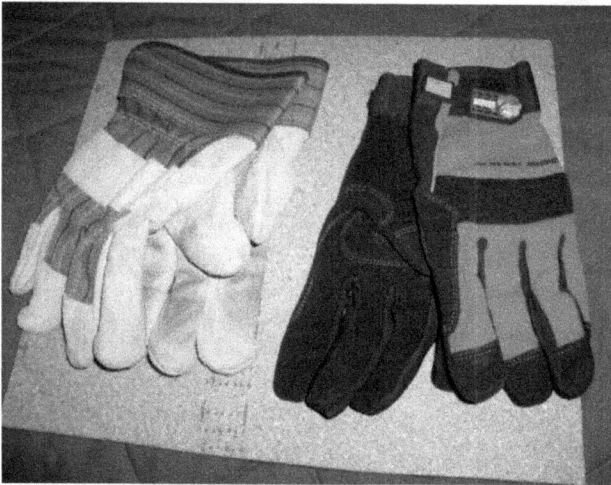

I clean the corrosion off the battery posts terminals two or three times a year. The battery posts terminals have a negative side (on the left) which is represented by a minus (-) sign, and a positive side (on the right) which is represented by a plus (+) sign. Corrosion builds up between the battery posts terminals and the connection points of the two cables that are connected to the terminals. The corrosion prevents a solid electrical contact for the battery to charge itself. A clean connection between the terminals and the connection points of the two cables will insure that the engine in your vehicle starts any time of the year.

I am going to tell you about an old mechanic's trick to keep your battery posts terminals from corroding. The trick is to use a pure petroleum jelly to coat the posts and the connection points for the two battery cables. You can find Vaseline and other pure petroleum jellies at drug stores or grocery stores like Wal-Mart stores in the health and beauty sections. The pure petroleum jelly prevents air from coming in contact with the posts and the connection points of the two battery cables. The exposure to air and the electrical charging and discharging of the battery causes the corrosion on the battery posts terminals and the connection points of the two cables. You have to remove the two battery cables before you clean the battery posts terminals and the connection points of the two cables. Use an open-end wrench or an adjustable wrench to loosen and remove the two battery cables from the battery posts terminals. The negative (-) cable is on the left side of the battery posts terminals. The positive (+) cable is on the right side of the battery posts terminals. If you have trouble removing the cables from the connection points, use a flat-tip screwdriver to lift the cables up .If the cables are too hard to remove, put the flat end of the flat-tip screwdriver under the connection points of each one of the cables to the battery posts terminals. Lift the cables up gently, and they will disconnect from the battery posts terminals. Next, use a tool called a *battery posts cleaner*. The *battery posts cleaner* cleans the battery posts terminals and the connection points of the two cables to the battery.

This tool uses stiff steel brushes to clean the surfaces of the battery posts and the connection points of the battery cables. The battery posts and the connection points of the two battery cables are made of lead or steel. It depends on the make, the model, and year of your vehicle.

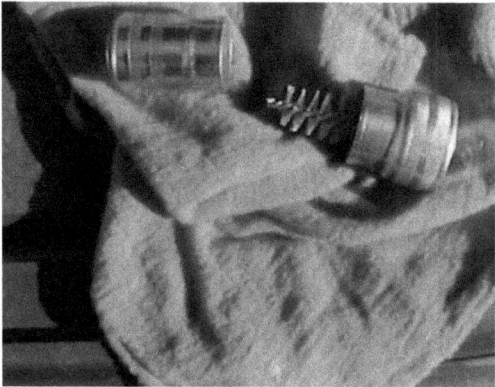

These are the parts to a *battery posts cleaner*. There is a removable cap on the top of the cleaning tool. (To remove the cap, twist the cap to the right; then, pull off the cap.) The cap protects the stiff steel brushes that clean the connection points of the two battery cables.

The bottom of the cleaning tool has a hole in it. The bottom of the cleaning tool is used to clean the battery posts. Do not stick your fingers inside the *battery posts cleaner*. You will cut your fingers. Wear the leather work gloves or general purpose mechanic's gloves.

Before you use the *battery posts cleaner* on the battery posts, you have to replace the cap. The cap is easy to replace. You have to push the cap back onto the steel brush holder. (Make sure the indentations on the steel brush holder and the cap match.) Turn the cap to the left until it locks in place.

Here is how to use the battery posts cleaning tool. Place the battery posts cleaning tool onto the left battery post terminal as shown in the photograph below. After placing the battery posts cleaning tool onto the negative (-) terminal, turn the cleaning tool two or three times to the left (with your hand) and then, two or three times to the right (with your hand). Now, pull the battery posts cleaning tool off the left battery post terminal. You should see a clean post terminal. If the post terminal is not clean, do it again. Clean the positive (+) terminal the same way. You cleaned the negative (-) terminal. Place the battery posts cleaning tool onto the right battery post terminal as shown in the photograph below. After placing the battery posts cleaning tool onto the positive (+) terminal, turn the cleaning tool two or three times to the left (with your hand) and then, two or three times to the right (with your hand) . Now, pull the battery posts cleaning tool off the right battery post terminal. You should see a clean post terminal. If the post terminal is not clean, do it again.

Clean around both battery posts terminals to get rid of the debris that accumulates after using the *battery posts cleaner*. Use a clean rag. When the surfaces look shiny and clean, then, you can use the pure petroleum jelly to cover the battery posts terminals.

Put some pure petroleum jelly on your index finger and start coating the battery post and the base of the post on the battery. Do this for the negative (-) and positive (+) battery posts terminals. (See the photograph below.)

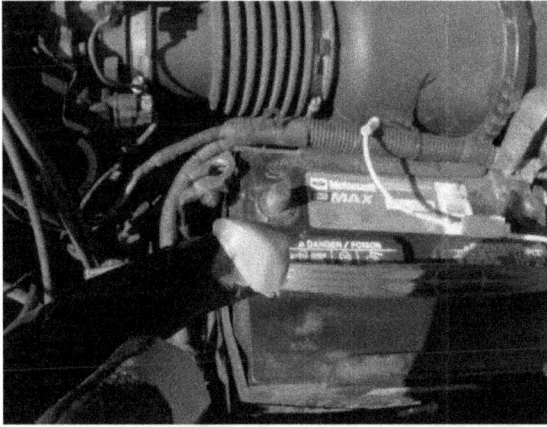

It is easy to clean the battery posts cables too. There is a negative (-) cable and a positive (+) cable. You need to remove the cap from the battery posts cleaning tool. Use the steel brushes to clean the cables. Start with the negative (-) cable. Insert the stiff steel brushes into the battery cable. (See the photograph below.) Grip the bottom part of the battery posts cleaning tool with your hand and turn the stiff steel brushes two or three times to the left and then, two or three times to the right. Make sure your fingers do not get stuck in the battery posts cleaning tool. It will cut your gloves and fingers. When you are finished cleaning the negative (-) cable, remove the battery posts cleaning tool and clean the debris off the cable with a clean rag. Next, apply pure petroleum jelly to the cleaned battery cable with your index finger. Put the jelly inside the two holes and on the top surface and bottom surface of the negative (-) cable. Clean the positive (+) cable the same way. Insert the stiff steel brushes into the battery cable (See the photograph below.) Grip the bottom part of the battery posts cleaning tool with your hand and turn the stiff steel brushes two or three times to the left and then, two or three times to the right. Make sure your fingers do not get stuck in the battery posts cleaning tool. It will cut your gloves and fingers. When you are finished clean the positive (+) cable, remove the battery posts cleaning tool and clean the debris off the cable with a clean rag. Next, apply the pure petroleum jelly to the cleaned battery cable with your index finger. Put the jelly inside the two holes and on the top surface and bottom surface of the positive (+) cable.

Now that the battery cables and the battery posts terminals are cleaned and coated with pure petroleum jelly you can put the battery cables back on the negative (-) and the positive (+) battery posts terminals. You need to tighten the bolts that secure the cables to the battery posts terminals. The pure petroleum jelly seals the battery cables and the battery posts terminals from the air that causes corrosion. Remove any jelly that is on other parts of the battery with a clean rag. If you take your time, you will not have a mess all over the battery. The picture below is a better photograph of the battery posts terminals and the battery cables coated with pure petroleum jelly and the two cables fully reconnected and secured to the battery posts terminals. By keeping the battery posts terminals and cables clean, your vehicle's battery will work to its full capacity and should last the rating marked on the battery.

Do not believe every mechanic's trick that they tell you. Never put a soft drink or any other type of liquid on a battery to clean it. I would never do this. Any sane knowledgeable mechanic would never do this. All liquids or oils with additives are conductors of electricity on a battery between the two battery posts terminals. The battery could "ground out" (or "short out") and cause an explosion. What does it mean to "ground out" or "short out" a battery? When electricity flows from the positive battery post terminal to the negative battery post terminal, the liquid or oil with additives between the posts terminals is a conductor. The charging and discharging from the battery causes hydrogen gas to form in the battery; the rapid discharging of the battery causes the battery to "ground out" or "short out." The rapid discharging between the negative (-) and positive (+) posts terminals builds up excessive heat from the rapid transfer of electricity between the battery posts terminals. Hydrogen gas will be building up more rapidly; it will **not** be able to escape from the vent caps. The battery could also explode and cause a vehicle fire. Do not put your tools between the battery posts terminals. The tools are conductors of electricity. The battery will "ground out" or "short out" if the tools come in contact with the battery posts terminals. I am prudent (wisely careful) and mindful (aware) of the dangers of mistreating a battery.

Use baking soda to neutralize the battery acid and to clean the battery. You can find baking soda in the baking section of a grocery store. Make sure you use a clean rag to clean the battery. Dispose of all the dirty rags in the garbage after working on your vehicle. Never wash or dry the dirty rags. The rags will cause your washer to have a foul smell, and your clothes will be ruined. The heat from the dryer could cause an explosion if the rags have any grease or acid residue on them.

Here is additional information about your vehicle's *electrical system* that works with your vehicle's battery. The *alternator is a .device in your vehicle that produces electricity to operate all of the electrical equipment in your vehicle. The alternator also charges the battery.* The battery stores the electrical power that the *alternator* produces, so the electrical equipment will operate longer in your vehicle. If your vehicle's *alternator fails or does not produce enough voltage for your vehicle's electrical system to operate,* your vehicle's *electrical system will start drawing electrical power from the battery while you are driving.* This will drain all the power stored in the battery and your vehicle will stop moving.

What devices and equipment make up the *electrical system* in your vehicle? The *brain box* "controls the operation of an engine and parts of the transmission" in today's vehicle's. The lighting equipment works with the help of the electricity produced from the *alternator* and stored in the battery. The lighting equipment includes *interior lights* and *exterior lights*, and accessory packages offered by the automotive companies work with the *alternator* and the stored electricity in the battery. For example, *heated seats, heated and powered mirrors, powered door locks, powered windows, a musical sound system package, a navigational system package, a powered roof package, electric fuel pump, and a clock are part of the electrical system.* Do not forget to reset the clock after you are done cleaning the battery posts terminals and the battery cables. When you disconnect the battery cables, the clock stops. Your owner's manual will show you how to reset the clock in your vehicle. Today, some vehicles have *electrical assisted steering* which uses an electrical device to help turn the steering wheel left or right. It saves gas because the steering mechanism is no longer driven by the gasoline engine, but is now driven by an electrical motor that works with the *alternator.* You also save money because an electrical device is being used to help you steer your vehicle. When the *alternator* fails or does not produce enough voltage, you will not have any power steering after the battery is drained of its stored electricity. You will have to steer your vehicle manually. There are more devices and equipment that rely on the *alternator* and the battery in your vehicle.

The *alternator* in my mini-van failed three times so far. I simply wore out the alternators because of the amount of mileage I drove my Ford Windstar. I thought I should pass this fact on to you.

Check the *alternator* at least once a year to find out if your vehicle has any problems with its alternator and battery. This is also a good way to check your vehicle's *electrical system* and to make sure it is working properly. You will have to take your vehicle to a mechanic's shop or a dealership. They have the tools to check the *electrical system* in your vehicle. You will have to ask them to check the *alternator* and the battery in your vehicle. They will be happy to do this for you.

Chapter 10

Light Check: Lights Anyone

I would like to start this chapter with this thought that the exterior and interior lighting system of your vehicle is just as important as any other part of your vehicle's operating systems. The lighting system of your vehicle needs to be checked to see if it is in good working order. The main exterior lights to be concerned about are the *headlights, turn signal lights, side marker lights, cornering lights, tail lights, brake lights, high-mounted brake lights, backup lights, and license plate lights*. These lights are the exterior lighting system on my Windstar.

Here is a brief explanation of the functions of the exterior lights on my Windstar. The *headlights* have a high and low beam. These are the main lights that help you see at night when you are driving. The headlights also help others to be able to see you at night. Never drive at night with your headlights turned "off." Always have your headlights turned "on" in bad weather conditions day or night.

Turn Signal Lights let people know if you are going to turn left or right in your vehicle. There are too many people that do not use the turn signals on their vehicles, and accidents are caused by not using turn signal lights. *Hazard Flashers* work with the turn signal lights, and are a safety feature in bad weather or when you have problems with your vehicle.

Cornering Lights work in conjunction with the turn signal lights. These lights are on the left and right side of the front bumper. They help other drivers and pedestrians to see whether you will turn your vehicle right or left.

Side Marker Lights are amber in color. They are behind the headlights on the left and right side of the vehicle. When you start up your vehicle, they are turned "on." They mark the sides of a vehicle like Cornering Lights and Turn Signal Lights for making turns.

Brake Lights tell other drivers when you are going to slow down or stop your vehicle. The fastest way to wear out your brakes is to follow too closely the vehicle in front of you. Riding your brakes is not the way to stop your vehicle from hitting the rear bumper of the vehicle in front of you. There should be one vehicle length (of your vehicle) for every *10* miles-per-hour that you drive on a highway or a road between you and the driver in front of you. If the posted speed limit is *50* miles-per-hour, you should be five vehicle lengths behind the vehicle in front of you. The vehicle lengths should be longer in bad weather conditions. Remember to drive slower in bad weather conditions.

High-mounted Brake Lights work in conjunction with the brake lights. This is a set of lights mounted on the back door or hatch door depending on the make and model of the vehicle. These lights can also be mounted in the back window or on the trunk of some vehicles. Your braking and stopping intentions are made more visible to other drivers with these lights.

Rear Tail Lights show the ends of the back of a vehicle. These lights work in conjunction with the turn signals and brake lights on your vehicle. Rear Tail Lights show other drivers that are behind you whether you are going to make a left or right turn, or slow down, or slow down and stop.

Backup Lights are located in the rear tail lights. These lights are turned "on" when you move your vehicle in reverse.

License Plate Lights are the most ignored lights until the police remind you by pulling you over to tell you that your license plate lights are out. You can avoid fines by making sure your exterior lights on your vehicle work.

Most vehicles have these exterior lights with some additional lights. Fog Lights and mirror-mounted turn signal lights are popular on some vehicles today.

I suggest that you do a light check at least once or twice a month on your vehicle to make sure you do not have any burned out exterior lights on your vehicle. The best time to do a light check is at night, after sunset, early morning, or on a cloudy day. During these times, you will be able to see which exterior lights are working and which exterior lights are not working.

It is easy to check the exterior lights of your vehicle. Begin by starting your vehicle and turning on all exterior lights. Turn "on" the headlights to low beam and turn "on" the left turn signals. Step out of your vehicle and walk around your vehicle from the driver's side door to the rear of the vehicle, and then, walk around from the rear to the front passenger side door of your vehicle checking all of the lights. When you are on the passenger side of your vehicle, walk around the front of the vehicle looking at the side marker lights and headlights while walking back to the driver's side door. Once inside your vehicle again, turn "on" the headlights to high beam and the right turn signals. Step out of your vehicle again, and look at the rear and front lights of your vehicle. Make sure the right turn signal lights and the high beam lights are working. To check the hazard flashers, just turn them "on," and step out of the vehicle to watch the flashing hazard lights in the rear and front of your vehicle.

Now check the brake lights, backup lights, and rear tail lights. Do all this at night with your vehicle's lights turned "on." The rear tail lights are turned "on" when the headlights are turned "on." You should see the red, rear tail lights in the side view mirrors and rear view mirror. Press down on the brake pedal with your foot until the brake lights are turned "on." Now, look in the side view mirrors and rear view mirror to see if the brake lights are "on." The brake lights should get brighter when you press on the brake pedal that indicates that your brake lights are working. You can get someone to help you by telling the person to stand in an area near the back of the vehicle. Tell the person to watch and to let you know when the brake lights are turned "on" as you press down the brake pedal. **Never** allow anyone to stand behind your vehicle when your vehicle is not in the park position on the steering column and your foot is on the brake pedal. Always keep the person in view when you look in your side view mirrors and rear view mirror.

Backup Lights are checked by putting your vehicle in the reverse position on the steering column and pressing your foot on the brake pedal. Keep your foot on the brake pedal. The vehicle will roll backwards if you do not keep your foot on the brake pedal. The person helping you could get hurt. You should see a white or clear color light turn "on" when you look in the side view mirrors and the rear view mirror. The backup lights should light up the rear of your vehicle.

The interior lights in my Windstar are the *dashboard lights, dome lights, cargo lights, map lights, door lights, and reading lights.*

Dashboard Lights are turned "on" at night with the headlights. This light illuminates the instrument panel of your vehicle. You will be able to see all of the gauges and warning lights. The dashboard lights are also background lights for the radio, clock, heating, and air conditioning panel.

Dome Lights illuminate the interior of a vehicle any time the doors of a vehicle are open. The dome lights are in the overhead ceiling of the front and middle area of my mini-van.

Door Lights are on the inside of the front doors and on the electric doors in the middle area of my mini-van where passengers sit. These lights help you see inside the vehicle when you open the doors.

In my mini-van, the *map lights* are on both sides of the dome light in the front of the vehicle. The map lights give you more lighting to read a map.

Reading Lights can be on the electric doors or in the overhead ceiling in the middle area of my mini-van. These lights help the passengers to read or see in the middle area of the mini-van when they are trying to find something.

The *cargo light* is in the back of my mini-van on the door. It helps light up the back of the mini-van, so that you can see to put things in the back or take things out of the back of the mini-van.

Interior lights are easy to check; just turn them "on" to see if they work once you are inside your vehicle.

If you have burned out lights on the exterior or inside the interior of your vehicle, your owner's manual can tell you the type of light bulbs you need for replacements. Some owners' manuals may tell you how to change all of the bulbs. The do-it-yourself books I mention in Chapter 6 can help you or you can go to a dealership or a mechanic's shop to get the light bulbs replaced.

Be careful with the headlight bulbs when you replace them. Why should you take special care to replace the headlight bulbs on your vehicle? The reason that you cannot touch the glass part of the headlight bulb is, Halogen light bulbs create a lot of heat when they light up. It does not matter whether the Halogen light bulbs are turned "on" low beam or high beam. Any *skin oils* or *any other types of oil* on the glass part of the Halogen light bulb could cause the light bulb to blow-out or stop working. (See the photograph below.) I am pointing to the glass tube that will blowout on a Halogen light bulb.

You can get *special gloves* to wear when replacing Halogen headlight bulbs at automotive discount stores. There are two types of gloves. One pair of gloves is a rubber Latex glove. (See the photograph below. The gloves on the left are the black rubber Latex gloves. The gloves on the right are rubber Nitrile gloves.) Both sets of gloves prevent skin oils and other types of oils from getting on the Halogen headlight bulbs. They come in many colors.

When you handle Halogen headlight bulbs **always** handle *the connector end* and *take your time replacing the bulbs to prevent bulb failure.* Not all vehicles have Halogen headlight bulbs. Be careful also when you are replacing the other light bulbs on the exterior and interior of your vehicle. You can avoid problems by taking your time.

I have a small collection of light bulbs for my Windstar mini-van. It is handy to have spare light bulbs when I need them and that includes spare headlight bulbs. I keep the bulbs in the glove box of my Windstar for emergencies. *It is cheaper to buy a*

headlight bulb than paying a ticket from a law enforcement officer for driving with one headlight not working.

There are two types of light bulb grease that you can use on the light bulb connections and in the light sockets for the lights on the exterior of your vehicle. I use *dielectrical grease to seal the light bulb connections* that are exposed to the elements like snow and rain. It is a thick grease that seals. The *dielectrical grease* is an *electrical conductor*. What is an *electrical conductor*? This means *electricity can go through the grease*. You need to put *dielectrical grease* on the light bulb connection and in the light socket before putting the bulb into the socket. You can also put *dielectrical grease* on exterior light bulbs that are still working.

The second type of grease is *bulb grease*. It is a thin grease for lubrication. *Bulb Grease also is an electrical conductor. Bulb Grease also prevents corrosion on the electrical contacts making it easier to replace light bulbs.* You need to put *bulb grease* on the light bulb connection and in the light socket before putting the bulb into the socket. You can also put *bulb grease* on exterior light bulbs that are still working.

You can find *dielectrical grease* and *bulb grease* in small packets at automotive discount stores. (See the photograph on the previous page. The brown packet on the left is the *bulb grease*. The orange packet on the right is *dielectrical grease*.)

I need to give you some brief information about *fuses*. There is a *fuse box* on all vehicles. The *fuse box* is located on the driver's side of the vehicle under the *dashboard* or behind the *dashboard*. Your owner's manual will show you where the *fuse box* is located. It will also show you the type of *fuses* needed for your vehicle. The *fuses* protect the electrical system in your vehicle from overloading with electricity. There are two main reasons why an electrical system fails. The first reason the electrical system fails on your vehicle is when an electrical part fails on your vehicle. For example, the reason the cigarette lighter fails is that the electrical fuse for the cigarette lighter has blown. The failed *fuse* protects the electrical system in your vehicle from damage. You will hear a popping sound. The second reason an electrical system fails is, the *fuse* has blown because of age. The electrical system in your vehicle that uses *fuses* will fail if a *fuse* blows out because of its age. You can buy *auto fuses* at an automotive discount store. The *fuses* come in a variety pack. The package has a variety of *fuses* ranging from *5 amps.* to *30 amps.*

What does *amperage* mean? *Amperage* is *the strength of the electrical current going through the electrical wires. Ampere is a unit of measurement for measuring electrical current. Ampere is abbreviated amp.*

The photograph below is a variety pack of *fuses* with a *combination fuse-puller and fuse-checker* for any vehicle. The following photographs shows you what a fuse looks like and how to use the combination fuse-puller and fuse checker tool

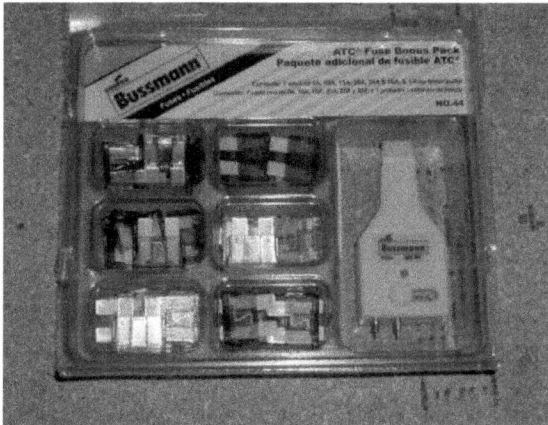

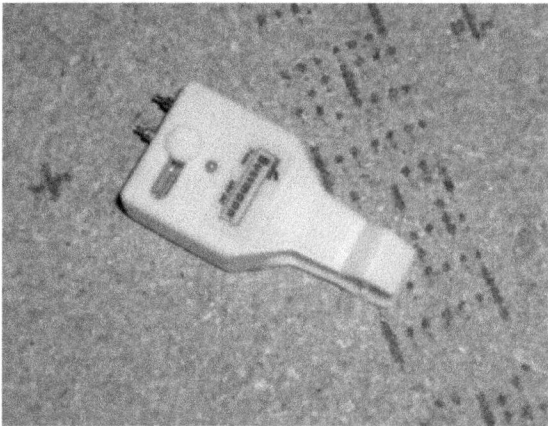

The photograph below shows you the different *amperages* of the fuses. The *amperages* range from *5 amps* to *30 amps*. The

amount of *amperage* is stamped on the colored end of the fuse at the bottom of the fuse.

The photograph above shows you a working fuse. Press the fuse-checker end of the tool onto the metal end of the fuse. When the green light comes "on," it means the fuse is working.

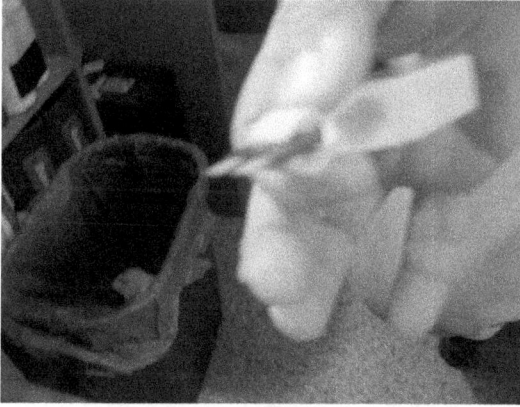

The photograph shows you the fuse-puller end of the tool. Use this end of the tool to pull a blown *fuse* from the *fuse box* in your vehicle.

Lights, camera, action—take the action to make the lights on the exterior and inside the interior of your vehicle operate longer. You will be glad you did.

Chapter 11

Wiper Blades: I Can See

Wiper Blades are on your vehicle's windshield for a reason. Wiper Blades help keep your vehicle's windshield clean during rain, snow, ice, and bug storms and the occasional four-wheeling mud adventure. *Wiper Blades* are *wear items that you replace* just like *tires, filters, and light bulbs*. When wiper blades *wear out from use*, they cannot do their job of keeping the windshield clean while you are driving. A good indication that wiper blades are worn is to do a simple test by using the electrical windshield washer system

106

on your vehicle. Observe how much fluid is being wiped off the windshield with each wiper blade pass over the windshield. Your vehicle's windshield should be clean without any view obstructions or windshield wiper fluid in the area where the wiper blades sweep over the windshield. A clean windshield gives you a better view of what is going on around you. If you see any type of fluid beads (moisture from the elements of nature, vehicle washings, or windshield wiper fluid) after the wiper blades pass over the windshield, this is another indicator that the wiper blades are worn.

I normally change my vehicle's wiper blades about twice-a-year or if I see that they are not cleaning the windshield adequately. I use replacement wiper blades from Ford or Motorcraft. The reason I use wiper blades from Ford or Motorcraft is that I know they fit my Ford Windstar mini-van. These brands of wiper blades have *wear indicators* on them. The *wear indicators* show me when the wiper blades are worn and need to be replaced. *The wear indicator is a black color dot on each wiper blade frame that changes to a yellow color dot when the wiper blades are worn and need to be replaced.* Ford and Motorcraft wiper blades are the best wiper blades I have used on my Windstar. You can use Ford and Motorcraft wiper blades on other American manufactures' cars, trucks, and vans. You have to go to a Ford dealership to get the wiper blades. The photograph below shows a worn wiper blade with a yellow color indicator dot. Not all brands of wiper blades have *wear indicators* on them

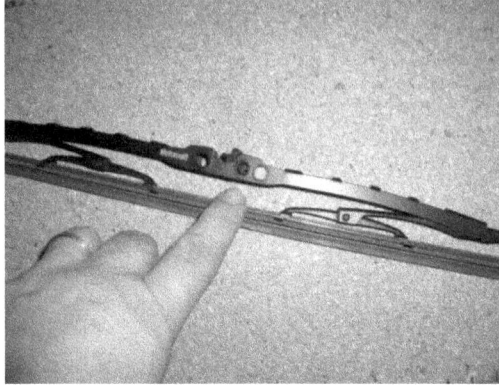

I would like to suggest that you take a good look at how the wiper blade attaches to the wiper blade arm before you take the wiper off the wiper blade arm. I would like to mention also that there are a lot of plastic parts holding the wiper blade to the wiper blade arm. *Plastic that is exposed to the outside elements of nature has a tendency to become brittle and can break very easily when you are trying to unhook the wiper blade from the wiper blade arm.*

Here is how you remove the windshield wiper blades. You have to take the old wiper blades off of your vehicle first. This is very easy to do. You start by lifting up the wiper blade arm slowly and bringing the wiper blade arm back toward you. The wiper blade arm has a spring on it to keep the arm in a standing position. Where the actual wiper blade and the wiper blade arm connect, there is a small plastic tab on the top side of the arm that you squeeze together to release the wipe blade from the wiper blade arm. Take your time when you are removing the wipe blade from the wiper blade arm. The plastic parts will break if you force them. Next, you slide the old wiper blade off the wiper blade arm. This

108

process of removing a windshield wiper blade will also work when you remove the rear view window wiper blade on your vehicle.

I take my time before I install the new wiper blades to make sure I have the correct plastic holders. Ford and Motorcraft have extra holders in the wiper blade packages for different makes and models of Ford vehicles, and they also have a generic holder for other manufacturers' vehicles. Compare the new wiper blades you just bought with the old ones on your vehicle to make sure they are the correct ones. I do this myself. I know I will not have any problems when I install the holders and wiper blades on my Windstar. Wiper Blades are just as important and essential as any other part on your vehicle. Buy good wiper blades that will do their job and last longer. Cheap wiper blades do not last and will not keep the windshield or rear view window clean.

This is how you replace a windshield wiper blade. First, attach the wipe blade holder to the wiper blade arm. Now, you can put the new wiper blade on the wiper blade arm. Hold the wiper blade arm in one hand and the new wiper blade in the other hand. Start at the top of the wipe blade arm. There is a hook at the top of the wiper blade arm. The hook has to be inside the new wiper blade in order to work. The hook attaches the new wiper blade to the holder on the wiper blade arm. Slide the new wiper blade up into the wiper blade arm until it snaps into place. You will hear a snapping sound when the new wiper blade is connected correctly to the wiper blade arm. You can use this process for replacing the rear view window wipe blade on your vehicle. Read the instructions on the wipe blade package if you have any questions.

For people who do not want to buy the wiper blades I use on my mini-van or cannot use them, there are many designs of wiper blades to fit all of the makes and models of vehicles that require wiper blades. The photograph below shows you one of the new wiper blades I use on my mini-van. You can see the black color dot on the new wiper blade frame.

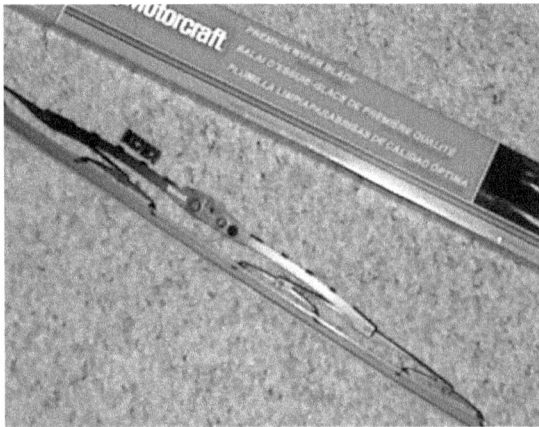

Let's use my Ford Windstar as an example for showing you the three different wiper blades I use on the front windshield and rear view window of my mini-van. The two front windshield wiper blades are different lengths. On the driver's side (or left side) of the vehicle, the wiper blade is twenty inches long. The wiper blade on the passenger's side (or right side) is sixteen inches long. The rear view window uses a twelve-inch long wiper blade. I change all of the wiper blades at the same time. This makes it easier for me to remember when I changed the wiper blades.

The last thing I want to mention is use your owner's manual to find the information you need about wiper blades for your vehicle.

One day you will thank me when it is raining very heavily and a semi-truck passes you and throws a huge wall of water on your windshield. The wall of water makes a driver temporarily blind (or has a blinding effect) to what is going on around him. Bad weather conditions can also obstruct your view while you are driving. New wiper blades can clean the windshield and rear view window, with the result that you can see to drive down the highway or road safely.

Chapter 12

Information About Keys

How much do you know about the keys to your vehicle? The subject about maintaining the keys to your vehicle is an interesting subject. Most people say, "Where are my keys?" They spend 30 to 40 minutes looking for them. The keys to your vehicle are only thought about when you are getting ready to walk out the door to go somewhere. Well, that has to change. Keys have changed quite a bit since the 1970's when I started driving a vehicle. Keys were designed to start up a vehicle, lock and unlock doors, open a trunk or hatchback door of a vehicle with one key. Today, keys to vehicles do more. You can start an engine remotely. You can lock and unlock doors, open and close doors, activate a panic alarm, and activate an anti-theft deterrent sound remotely because of a programmed, electronic chip in the *key remote transmitter*. The photograph at the beginning of the chapter shows you two sets of keys with remote transmitters. Some key remote transmitter designs are separated from the keys and activated with buttons. Other key designs combine the *key remote transmitter* within the key itself. This is called an *integrated key*. The *key remote transmitter* is still activated with buttons on the key. You can use the information I give you for both remote transmitter designs. I will be using my Ford Windstar key remote transmitter as a reference for teaching you in this chapter.

Let's start with the owner's manual of your vehicle. It should tell you how to replace the electronic battery inside your vehicle's key remote transmitter and the type of replacement electronic battery you need to buy. I change the electronic battery about every two years. Think about how many times you use the electronic key to remotely start or access your vehicle. Constant use of the buttons on an *integrated key* or on a *key remote transmitter design* causes the electronic batteries to work more; therefore, they need to be replaced more often. You do not want the electronic *integrated key* or a *key remote transmitter* to fail in an emergency. It is a good idea to have spare electronic batteries for the keys to your vehicle when you need them.

113

The photograph above is the replacement batteries is for my Ford Windstar mini-van's key remote transmitter. The electronic battery number can be seen in the upper left corner on the package. If the owner's manual of your vehicle does not state the electronic battery number, then, the easiest way to find the number is by opening the *integrated key* or the *key remote transmitter*. Look on the backside of the battery to locate the electronic battery number and the manufacturer of the battery.

The photograph on the next page shows you the inside of my Ford Windstar key remote transmitter. You can see the battery that fits inside the key remote transmitter housing. To open the key remote transmitter housing, you have to remove all of the keys from the key ring and the key ring that is attached to the *key remote transmitter*. You can use a coin on the top outer edge where the key ring attaches to the remote transmitter. A dime will work. Place the dime between the top outer edges of the *key remote transmitter (or an integrated key)*. Move the dime from side to side using a slow steady force until the *key remote transmitter (or the*

114

integrated key) opens. You will hear a popping sound. Once the *key remote transmitter* pops open, you can separate the housing covers with your hands. The same procedure can be done for an *integrated key*. The photograph below shows the two halves of the key remote transmitter housing. The plastic key remote transmitter will break if you are not careful. *Replacing* a broken *key remote transmitter* could cost you *thirty dollars to one hundred dollars* for a replacement transmitter, *and the cost of programing the transmitter to work with your vehicle*. The programming cost could be as high as *fifty dollars to seventy dollars* at a dealership. Your owner's manual can show you how to program a new key remote transmitter yourself.

You must have a *spare key* if you want to avoid locking yourself out of your vehicle. The *spare key* is only used to open the door on the driver's side to give you access to the inside of your vehicle. The spare key **will not** start your vehicle. Today, the cost of making a *spare key* for *opening the driver's side door* of your vehicle is about *two dollars to three dollars*. **Never** put your *spare*

key in a magnetic box that attaches underneath your vehicle's bumpers. This is an open door for a thief to steal items from your vehicle. The magnetic box could fall off your vehicle while you are driving. You can carry the *spare key* in your wallet, purse, or on a chain. I carry a spare key in my wallet. My *spare key* has saved me money. I did not have to call a locksmith to unlock my mini-van. The *spare key* helps people who do not have remote access packages on their vehicles. For example, *Onstar,* a subscriber service of the General Motors Corporation, has remote access to a vehicle if you get locked out of your vehicle.

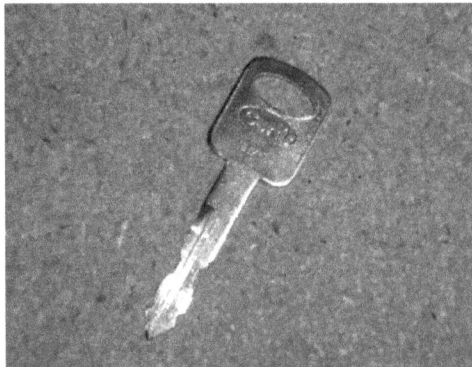

Ignition Keys have become more expensive to replace if you lose them. It is not uncommon to pay *thirty dollars to forty dollars* for a replacement ignition key. There is an anti-theft deterrent electronic chip mounted inside the plastic part of the *ignition key.* In the photograph below along the edge of the *ignition key,* you can see a spot on the corner of the key that looks like a rectangle. This where the anti-theft deterrent electronic chip is located in the ignition key. The chip is on the inside of the *ignition key.* When you buy a replacement ignition key, you will have to program the key to your vehicle which enables the new key to start your vehicle.

Your owner's manual will tell you how to program a new key to your vehicle, or you can have a dealership do it for you. The *programming cost* could be as high as *fifty dollars to seventy dollars* at a dealership. The best place to get an *ignition key,* a *key remote transmitter, integrated key,* and a *spare key* to access the driver's side door is from a dealership that sells the vehicle you are driving.

Chapter 13

Be Prepared

This chapter will teach you some of the basic things you need to carry inside your vehicle in emergencies or when you need to make quick repairs or replacement of things on your vehicle. Below is a list that will help you equip your vehicle with the necessary things you need for driving year-around and in any type of weather. Some of these things I carry in my Ford Windstar all the time. You can change the list according to your needs and where you live.

The basic be prepared list for your vehicle:

A set of *jumper cables* are used to charge up a vehicle's battery, but you need another person with a vehicle to charge your vehicle's battery. Your owner's manual should show you how to use Jumper Cables.

A *foldable, snow shovel* is handy in snowy and muddy weather conditions for digging yourself out when your vehicle's tires have no traction.

A spare *serpentine drive belt* or *V-drive belt* in case the one on your vehicle breaks. You can change the drive belt yourself with a *breaker bar* if that is the only thing that breaks on your vehicle. Most tow truck drivers carry a *breaker bar*.

A *breaker bar* is used to move the drive belt tensioner pulley to allow you to put on a new drive belt. The *breaker bar* is also used to loosen hard to turn bolts and nuts. You can buy a *breaker bar* at automotive discount stores.

A *12-volt, portable, air compressor* is used to inflate flat tires on the road or use the air compressor after you use a *tire repair kit*.

A *tire repair kit* is used to repair a flat tire that has been punctured.

Buy spares *fuses* and spare *exterior* and *interior light bulbs*.

Buy a spare *headlight bulb*.

Basic *hand tools*: You need two screwdrivers (a flat screwdriver and a Philips or Crosspoint screwdriver). A pair of pliers (slip joint pliers) and an adjustable wrench (an 8-inch or 10-inch length wrench).

A *tool box* to carry the things you need for your vehicle.

A *6-volt flashlight* last longer.

Tire gauges like the Dial Tire Gauge and the Pen Tire Gauge in Chapter 2.

Shop Rags to clean things before you put them back together.

Emergency Tow Ropes come in two sizes ten feet to fifteen feet for mid-size vehicles and up to thirty feet for larger vehicles. A *tow chain* can be used for larger vehicles. A *tow chain* measures ten feet to thirty feet in length.

Battery posts cleaner tool and a container of *pure petroleum jelly*.

Get a *blanket* and *pillow* for emergencies and long trips.

Use an *ice scraper* with a brush on it. It will help clean your windshield better in the winter. Use the *ice scraper* to clean the other windows and mirrors on the outside of your vehicle too.

A *bag of sand* can be used for extra weight in the back of a vehicle during rainy, snowy, and icy weather. The extra weight helps to increase the traction of the tires on your vehicle during bad weather conditions. The extra weight can help the rear end of a small vehicle from sliding out of control and losing traction during bad weather.

A spare *one-gallon gasoline container* will get your vehicle to the gas pumps after you walk to the gas station to fill the can. The cans come in plastic or metal. **Do not leave a filled gasoline container in your vehicle. The spilled gas, the heat, and the fumes around a lit cigarette could cause an explosion.**

Flares are used at night for lights when you and your vehicle are broken down on the side of the road.

Emergency Roadside Triangles do the same thing as *flares*, and they are safer to use.

An *Emergency First-Aid Kit* can be personalized according to your medical condition or needs in an emergency.

Make sure your *cellphone* is able to pick up a signal in the area where you are stranded.

The owner's manual will give you specific information about parts, fuses, lights, and light bulbs for the exterior and inside the interior of your vehicle. It will also show you how to use jumper cables. I have used my jumper cables and flashlight many times to help others in need. I carry one quart of 5W-30 oil for my Ford Windstar. I also carry one quart of Mercon V (Mercon 5) transmission fluid and a bottle of power steering fluid. (See the photograph below.) You can buy the things on this list at automotive discount stores, dealerships, mechanics' shops, tire shops, department stores, and grocery stores.

Part of being prepared is doing regular maintenance checks on your vehicle. It is a good idea for you to have yourself and your vehicle ready for most situations and emergencies.

Chapter 14

Driving Habits: Self-Improvement

I want to stress the importance of safe driving habits. Any time you get inside your vehicle to drive, you need to remember these two important facts. *Driving is a privilege and not a right. You can lose your driving privilege by driving recklessly and by using unsafe driving habits.* Why do I say this? The state that issued your driver's license gave you the driver's license because you passed a written test and a practical driving test. The state assumes that you will remember what you have learned in the driver's manual, and that you will drive safely with your vehicle.

Multi-tasking is great in your home or office but not in a moving vehicle when you are the driver of that vehicle. All of your attention needs to be focused on who or what is in front of you, behind you, and around you while you are driving. Many states in the United States of America and some of its territories have laws banning hand-held cellphone use and texting while driving. You can go to the following website to see which states have passed laws banning hand-held cellphone use and texting while driving. Http://www.ghsa.org/thml/stateinfo/laws/cellphone_laws.html. Multi-tasking does not mean you are a competent and safe driver. When you are driving be alert and aware, there are other drivers on the same road you are on trying to go somewhere like you.

We are living in a time with many computerized technological advances. Vehicles have not learned to operate themselves. The automotive industry is working on it. A vehicle still needs a competent, alert driver to operate it. *Your state of mind is important when you are driving.* Your first concern should be driving safely and obeying posted road signs, traffic signals, and painted lines on a road or highway. Your second concern should be not allowing anything to distract or deter your attention from driving. All of the following distractions and deterrents can be done before you start your vehicle: making telephone calls and cellphone calls, texting, playing with compact disks and the radio, adjusting mirrors, seats, and seatbelts. I pull over onto the shoulder of the road or find a suitable place to take a telephone call or to make a telephone call. A hands-free telephone system is nice to have in your vehicle, but it should only be used in emergencies while you are driving.

No one else is in control of the vehicle except you. Anything can happen while you are driving. Situations can change very fast.in city or highway driving and even on a country road. One of the vehicles I owned in the past was hit from behind because a dog sitting in the front passenger seat of the vehicle behind me jumped into the front seat with the driver and momentarily distracted the driver. The dog prevented the driver from stopping when I was braking for a traffic signal that was about to turn red, which caused me to stop my vehicle. The dog's owner hit the back of my vehicle. If your animal is not properly trained for riding in a vehicle or not in a cage, do not allow the animal to ride with you inside your vehicle. Be selective on who you will allow to ride with you in your vehicle. Do not allow family, friends, or your acquaintances in your vehicle who like to play jokes on the driver. Excessive talking to the driver, yelling, and fighting of the passengers in your vehicle is a definite distraction to the driver of a vehicle. Teach your children to have good manners while they are riding in a vehicle.

In bad weather conditions, you should be more aware of what is going on around you because the roads can be unsafe and hazardous for driving. I would like to give you a simple rule to think about. *Every 10 miles-per-hour that you are driving in your vehicle is equal to one length of your vehicle.* There should be one vehicle length for every *10* miles-per-hour that you drive on a highway or road between you and the driver in front of you. If the posted speed limit is *70* miles-per-hour, the minimum distance required to stop is seven lengths of your vehicle on dry pavement. You should be seven vehicle lengths behind the vehicle in front of you. The vehicle lengths should be longer in bad weather conditions. Reduce the speed of your vehicle in bad weather conditions. For example, if you are driving at a speed of 50 miles-per-hour on a wet road surface, the distance between your vehicle and the vehicle in front of you should be two or three times the length of your vehicle. Ten to fifteen vehicle lengths of your vehicle is the minimum distance required to give you plenty of space to stop your vehicle in rainy, snowy, or icy weather. Use your head; if you need to drive slower, do it. The longer the distance between your vehicle and the vehicle in front of you the safer you are.

When you are driving, always be in the habit of *scanning the road ahead and around you. Look for traffic that can become a problem.* The left lane drivers use the left lane for driving instead of passing vehicles. They do not move to the right lane to allow others to pass them. Pack drivers are too many drivers that drive to close together on a road or highway. There is no place to go if one of the vehicles in the pack has vehicle trouble. When I drive, I always look for an *escape route* in case of trouble. An *escape route* might be the shoulder of the road or any place away from an accident or trouble. I do not want to make the situation worse for everyone else. Keep your distance between you and the driver in front of you. Stay in the right lane or the lane furthest to the right when passing over a hills or going around curves. I have had people pass me on a two-lane road over a hill, and they ended up in the lane I was driving in. I was driving in the opposite direction. The driver almost hit me head-on. He could not wait for the slow driver in front of him. If this circumstance happens to you, drive your vehicle to the right shoulder of the road. Get off the road. Sometimes there is no *escape route*. I have been hit broadside by a driver in a vehicle on a highway who was trying to get away from the police that were chasing him. I could not avoid that accident.

Who is an unsafe driver? A person that is an aggressive, reckless, and inconsiderate driver to others is an unsafe and unstable driver. What causes a person's mind to be in a state that makes the person an unsafe and unstable driver? Alcohol could be one reason for reckless unsafe driving. Most of the states in the United States of America have open container laws that prohibit the drinking of alcohol while driving. A designated driver, taking away a person's keys, calling a taxi cab for a drunk person, or staying home when you drink the alcohol are good preventive measures to stop drunk driving. Some states have included drinking any type of beverage and eating any type of food while driving as part of their open container laws. A few cities in the United States of America also have this law. A drunk driver drove by me once on my way to work in the early morning hours. He drove by me smiling and he raised his beer bottle and sped away. I made a call to the police to alert them about his reckless driving. You can do a search on line to find the open container laws for the city, state, or country you live in and the penalties for violating these laws. Do not get in a vehicle with a drunk driver.

Drugs are another reason for a person's aggressive and reckless behavior while driving. The warning labels on prescribed medications are there to prevent driving fatalities and machinery accidents. Illegal drugs and even legalized marijuana impair a

127

person's judgment while they are driving. Never get in a vehicle with a person on illegal drugs or marijuana.

Emotional or mental stress can also impair a person's judgment while they are driving. Ask someone to drive you where you need to go or take a taxi cab or a bus. Get medical, legal, or spiritual help to get you through the mental or emotional stress that causes distracted driving.

A person that drives while he is tired and sleepy is a risk to himself and others. Pull over to the side of the road in an area where it is safe or go to a rest stop, truck stop, or a hotel, or motel if you are traveling long distances. Get the sleep you need. When you are traveling with friends or family, share the driving with them. A job with a crazy work schedule all the time or staying out late at night and going to work the next morning very tired is bad for your health. These situations are other causes for distracted driving. No one is invincible. Your body will naturally shutdown when it needs sleep. Some people have a rare sleeping disorder that never allows them to sleep. This disorder eventually kills the person. It is not narcolepsy or insomnia which are other sleeping disorders. Opening a window and a loud radio will not stop you from falling asleep. If you spot unsafe, reckless, aggressive, inconsiderate drivers, make a telephone call to police when you are able to do so safely. Tell them where you have seen an unsafe driver and describe the vehicle.

Make sure you are dressed, shaved, and have put on your makeup before you leave your home and get in your vehicle to drive. If you are late for work or an appointment, get up earlier or make a schedule, so you have time to take care of yourself. Do not dress your children in a vehicle while you are driving. *Remember unsafe driving is a danger to you and everyone on the road.* All of these distractions can be prevented by the driver of the vehicle. You will be saving your life and the lives of others on the road.

Road construction areas are something to be aware of while driving. There is a lot of activity in road conduction zones. Road Construction Workers are walking on the road and heavy equipment is being used to repair or build roads, highways, and bridges. There are posted detour signs to help you drive around the construction area. When there are no detour signs, then, you must obey the reduced speed limit signs to help save the workers lives and your own. I slow down and am very watchful in these areas until I have cleared the road construction area. The posted and reduced speed limits in construction zones are the maximum speed you can drive through a construction zone. In some states, there are posted signs in road construction zones telling divers the large fines and the maximum sentence for imprisonment drivers will receive for disobeying road construction zone signs and causing an accident or road worker fatality.

A good thing to do is always be in the habit of scanning the rear view and side view mirrors to check traffic if you are planning a lane change. You cannot wear out mirrors by looking in the mirrors when you are driving. By looking in the mirrors, you might see an ambulance, fire truck, or police vehicle coming toward you. If it is possible, get into another lane when you hear the siren of an ambulance, fire truck, or police vehicle. They are trying to save lives and property. Time is crucial for saving a person's life. If these emergency vehicles are behind you, pull over to the side of the road when you can do it safely and let them pass you. Some states have laws that tell the driver to pull over to the side of the road or make a lane change when they see and hear an emergency vehicle coming their way. Next time, it might be you who needs an ambulance, policeman, or fireman. Motorcycles are smaller vehicles. Motorcycles are hard to see; they can be in the blind spot of your vehicle. Always be watchful for them during good weather and rainy weather. Mopeds, bicycles, and pedestrians are other things to be alert and watchful fur when you are driving. Learn to be a defensive driver.

Who is a defensive driver? A person that is alert, watchful, anticipates trouble, and is prepared to take the correct, action to avoid trouble. The tips you learn in this chapter will help you be a defensive driver. There are many videos on the internet that can teach you about defensive driving.

Speed Limit Signs are part of being a defensive driver. Posted speed limit signs are the *maximum speed that can be safely driven on dry surface roads in specific areas*. The speed limit sign above says, "Begin" at the top of the posted speed limit. It tells you "where the *45* mile-per-hour zone starts." *It pays to read all posted signs to prevent problems while driving.*

Your driving habits will affect your vehicle's gas mileage and how much wear and tear your vehicle will endure while you are driving. Heavy brake use, excessive fast starts, or hard clutch shifting can play havoc on your vehicle. Use your seat belt it may save your life. Happy driving wherever you go. Be an alert, aware, and a safe defensive driver.

Chapter 15

Finding a Good Repair Shop

Vehicles built before 1975 were easy to work on. If a person had good mechanical skills and a basic tool set, he could work on his own vehicle. In the year 1975, the United States Congress passed CAFÉ Standard Regulations for the automotive industry. CAFÉ Standards is an acronym for Corporate Average Fuel Economy Standards. The regulations were passed in another bill by the United States Congress called the Energy Policy and Conservation Act of 1975. The Arabian oil embargo of 1973 started the beginning of the nightmare for excessive regulations for the automotive industry. The CAFÉ Standard Regulations are regulated by the National Highway Traffic Safety Administration. The EPA or the Environmental Protection Agency did the testing of vehicles to make sure the vehicles being built could meet the CAFÉ Standards. These regulations forced the automotive industry to build more fuel efficient vehicles or face heavy fines. The idea was to get more fuel mileage out of a gallon of gasoline. The result is vehicles today come with computerized brain boxes. The *brain box controls the entire engine functions of a vehicle. These functions include the timing or firing order of the spark plugs, fuel metering, idling of the engine, transmission shifting, exhaust emissions, and many other functions too numerous to list.* The automotive industry is required to make vehicles compliant with the United States government CAFÉ Standard Regulations for fuel mileage and emission standards for vehicles being built and sold in this country.

Working on vehicles that are controlled by computers is not an easy thing to do. The owners of computerized vehicles may have limited mechanical knowledge or no mechanical knowledge of computerized vehicles. Therefore, finding a capable and knowledgeable repair shop is very crucial. It is like finding a reputable doctor or hair stylist that you can trust to work on you. In other words, you need to find a repair shop that you can truly trust to work on your vehicle. One that does quality work with a guarantee and a warranty for their work and for the parts they install on your vehicle.

133

I bought my Ford Windstar mini-van in the year 2000. I am the original owner of my mini-van. Over the years, I have used many repair shops. The repair shops ranged from small repair shops to large Ford dealerships. The costs of doing repairs on your vehicle should not be the only criteria for choosing a repair shop. The costs of repairs for your vehicle should be considered when you are trying to find a reputable repair shop. A repair shop that quotes you a price for parts and labor and estimates the time needed to do the work and does the work in a timely manner cares about its customers. The costs should not be excessive. Keep in mind that the total costs of the repairs will not be known to the mechanics until you bring your vehicle into the repair shop. A reputable mechanic will give you reasonable estimated costs for labor and parts. Some repair shops have courtesy vehicles for their customers to use if their vehicles are going to be in the repair shop for an extended period of time.

The photograph above is a picture of the Don Meyer Ford and Lincoln repair shop in Greensburg, Indiana. They have very capable and knowledgeable mechanics that do quality work, and the work is guaranteed. The Ford Motor Company guarantees and warranties their Ford and Ford Motorcraft parts too.

One way to reduce the costs of a repair bill is to buy the parts ahead of time before you take your vehicle into the repair shop. I do this myself. I know parts will eventually need to be replaced; I buy the parts ahead of time. I make an appointment, as soon as possible to do the repairs. The following is a list of the parts I buy before getting repair work done: a water pump, brakes, ball joints, wheel bearings, belts, etc. A reputable repair shop will do a vehicle inspection report on a vehicle before they do any service work on a vehicle. Don Meyer Ford and Lincoln do this for their customers. They have a detailed inspection sheet they use to evaluate the status of a vehicle. The Vehicle Inspection Report Sheet is at the end of this chapter.

The Vehicle Inspection Report Sheet is used at Don Meyer Ford and Lincoln when *service work or scheduled maintenance* is being done on a vehicle. What is s*ervice work* or *scheduled maintenance* on a vehicle? *Service Work or Scheduled Maintenance is routine maintenance on a vehicle. Rotating tires, oil and filter changes, replacing belts, and replacing wiper blades are considered service work or scheduled maintenance.* Different repair shops will use one of these two names for *service work.* What is *repair work? Repair Work is replacing specific items on a vehicle that wear out over a specific period of time. For example, worn brakes and a worn water pump have to be replaced over time.* The Vehicle Inspection Report Sheet gives you a detailed explanation of the status of your brakes, tires, battery, lights, windshield, belts, hoses, steering system, the exhaust system, and the transmission drive axle. The report gives you a general overview of your vehicle's condition. At Don Meyer Ford and Lincoln, they take the time to explain their findings on the report before they do any *service work* or *repair work*.

Today's vehicles are computerized. You need to take your computerized vehicle to a reputable repair shop or dealership when it is having problems. They can do a computerized diagnostic review on your vehicle. The mechanics use a book of codes to specifically pinpoint the computerized problem with your vehicle. The computerized diagnostic review cost about *fifty dollars to one hundred dollars* depending on the type of diagnostic equipment the repair shops use. Make sure you get a diagnostic review done on your vehicle before specific repair work is done. This will save you time and money. The service managers at Don Meyer Ford and Lincoln explain in detail the computerized codes to me before they do any *repair work* on my Ford Windstar.

I would like to mention that a majority of the repair work on my Ford Windstar was done by a local area Ford dealership, Don Meyer Ford and Lincoln, located in Greensburg, Indiana. This Ford dealership did more for me than other Ford dealerships that I have used for *repair work* and *service work* in the past. *They took the time to show me how to prevent other problems which saved me money.* For example, when my mini-van had unusual tire wear patterns on the front tires, the mechanics said my vehicle might need a front-end alignment. The front-end alignment saved the tires from being replaced prematurely. When I had to replace the transaxle on my mini-van, the mechanics noticed that the ball joints on my vehicle's lower suspension needed to be replaced.

The mechanics said it would be cheaper to replace the ball joints while they were replacing the transaxle. I would pay more money if I had to make another appointment to replace only the ball joints. I stayed with this dealership because they do a great job of keeping my Windstar operating. They have a helpful parts department and knowledgeable service managers. That helped me learn more about my Windstar. The mechanics at the dealership have awards from Ford showing their knowledge and mechanical skills.

Dan Scheidler is one of the three *Service Mangers* at Don Meyer Ford and Lincoln. He explains in detail the Vehicle Inspection Report Sheet to a customer. He also answers any questions a customer may have about the *repair work* or *service work* being done on their vehicle. Dan and the other Service Managers discuss future problem areas a customer will have on their vehicle. These problems are concerns that the mechanics have found while they are working on a customer's vehicle. Don Meyer Ford and Lincoln has a very good repair and service department. I use them as a comparative tool to compare other repair shops' and service shops' work, prices, warranties, and labor costs.

Throughout this chapter I have tried to give you specific criteria to use in finding a reputable repair and service shop for your vehicle. It depends on the type of work your vehicle will need.

You can ask your family, friends, and acquaintances who they trust and go to for serving and repairing their vehicles. The Better Business Bureau in your area might be able to tell you whether a business is reputable and if there are complainants filed against the business. Not all businesses are listed with the Better Business Bureau. You can use a telephone book and look for local repair shops and dealerships in the area that work on the make and model of the vehicle that you own. Explain to the mechanic as best you can what is wrong with your vehicle. If the repair shop cannot help you, they may refer you to a repair shop that specializes in the work that needs to be done on your vehicle.

What questions should you ask once you have found a repair shop? What are the costs for parts? What are the costs for labor? Are there warranties on the parts being put in my vehicle? Are there warranties on the labor the mechanics have done on my vehicle? How long will it take to make the repairs on my vehicle? If my vehicle has to be in the repair shop over night or a few days, does the repair shop have a courtesy loaner vehicle? Call at least two or three repair shops or service shops to compare pricing for labor and parts, warranties, and what type of work each shop is willing and capable of doing on your vehicle. Keep in mind that a repair shop will not know what the exact costs of a repair on your vehicle will be until they actually look at your vehicle and start working on it.

Vehicle Inspection Report Sheet

Chapter 16

The End of the Road

You and I are at the end of the book, but looking forward I am trusting you to take the knowledge I have taught you and apply it. I do not want to give you nightmares about the owner's manual for your vehicle. It can answer many of the questions about your vehicle. I cannot say I know it all, but I have tried to teach you the basics of how to keep your vehicle running longer. I kept my 2000 Ford Windstar mini-van operating for fourteen years. When this chapter was written, my mini-van had 458,967.3 miles on the odometer. My mini-van still runs, but the odometer and the speedometer have failed. I know it can be done; I have proven it;

all it takes is the desire and commitment to make it happen. I had fun putting this vehicle knowledge in print for others. All I can say at this point is have fun with your vehicle, and do not be afraid to learn something new. Do not be afraid to ask questions when you do not know what to do. Happy driving!

March 5, 2009

Dear Ford:

    I want to thank you for building the Windstar mini-van. Mine is a 2000 model and to this date my Ford Windstar has 363,984.5 miles. It is still racking up more miles every day on the original motor and transaxle. I have replaced tires, brakes, and other things that wear out. I bought my Ford Windstar on April 10, 2000 with 12 miles on it. I am writing a book entitled *Life After 100,000 Miles* or *How to Keep Your Car Going Longer*. The book will be for those who want to keep their cars going longer, and it will show them how to do it. Simple things that anyone can do to keep their cars longer. I have enclosed pictures of my Windstar to show you what it looks like with all the miles I have driven on my Ford.

    I plan on buying another Ford car, either a Focus or Taurus, but I am thinking of the new Ford Fiesta that is coming out soon. I have had very good luck with Fords and I plan to buy more in the future. Most of all, I plan on keeping my Ford Windstar past 425,000 miles, which I think it still can do with the original motor and transaxle.

    I enjoy my mini-van a lot to this day, and I use it to haul everything; from carrying lumber, to carrying my family 4, and soon, I will be pulling a small fold up trailer that I own. I just want to tell Ford thanks for building my Windstar mini-van, and it is still a great vehicle even with all the miles on it!

Sincerely,

Theodore R. Hansen

143

These are the photographs I sent to the Ford Motor Company on March 5, 2009.

145

Reply from Ford

Ford Customer Service Division

P. O. Box 6248
Dearborn, MI 48126

March 19, 2009

Theodore Hansen
14375 N. Morris Washington St.
Batesville, IN 47006-8281

Case # 1448170779

Dear Theodore Hansen,

Thank you for contacting us.

We are happy to know that your vehicle has provided such good service over the many miles it has been driven.

We have been producing automobiles since 1903; we are proud of our past, but each year becomes a new challenge. From management to the assembly lines, "Quality is Job 1" remains our continuing objective. Your satisfaction with your vehicle is proof that our efforts have paid off. Of course, your good driving habits and proper maintenance have helped greatly.

Thank you for your continuing loyalty and the time you have taken to share your experience with us. We are pleased to have you in the family of satisfied owners.

Sincerely,

Teresa Wesley
Ford Motor Company
Customer Relationship Center

About the Author

This is the hardest part to write because I am talking about me, the author. I rebuilt a 1963 Pontiac Tempest convertible from the ground up. This sparked my interest and desire to become a mechanic. I learned the basics of a mechanic's career and gained much knowledge about the operation of a car. I also worked on other vehicles that I have owned in the past to gain even more knowledge than the basics. My desire to learn about mechanical things has motivated me. I am a licensed aircraft mechanic. I got my A&P license in the United States Air Force. I still work as a draftsman. I think with a creative mind that never sleeps. My mind always thinks of the possibilities of what it could achieve. This book was an achievement for me. I spent five years writing this book.